S/O - LTL BIOENG
No Info

MICROSCOPY HANDBOOKS 29

Flow Cytometry

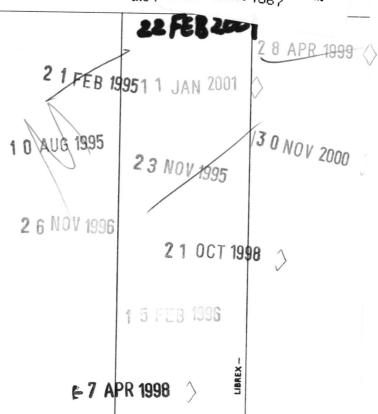

Royal Microscopical Society MICROSCOPY HANDBOOKS

Flow Cytometry

M.G. Ormerod

34 Wray Park Road, Reigate, Surrey RH2 0DE, UK

βIOS
SCIENTIFIC
PUBLISHERS

In association with the Royal Microscopical Society

First published in the United Kingdom 1994 by
BIOS Scientific Publishers Limited,
St Thomas House, Becket Street, Oxford, OX1 1SJ
Tel. 0865 726286; Fax 0865 246823

A CIP catalogue for this book is available from the British Library.

ISBN 1 872748 39 2

Typeset by AMA Graphics Ltd, Preston, UK
Printed by Information Press Ltd, Oxford, UK

Preface

This handbook is based on the Flow Cytometry Courses organized in the Department of Pathology, University of Cambridge, by the Royal Microscopical Society. Its aim is to introduce workers new to the field to the concepts underlying the instrumentation and uses of flow cytometry. I hope that the reader will also obtain some appreciation of the wide range of applications of this exciting technology.

Flow cytometry has been used to study a wide range of organisms, including plant cells, yeasts and bacteria. The large majority of publications have been on mammalian cells and, partly for this reason and partly because of the limitations of my own experience, this handbook deals exclusively with mammalian systems.

Practical protocols have not been given in this short handbook since these have been covered elsewhere by several authors, including myself (Ormerod, M.G., ed. (1994) *Flow Cytometry. A Practical Approach*, 2nd edn. IRL Press, Oxford).

I would like to thank the many colleagues who have worked with me over the years and, in particular, Mrs Jenny Titley, who recorded much of the data shown in the handbook. The contributions of other colleagues who have supplied me with data are acknowledged in the text.

I would also like to thank Drs Mike Ayliffe, Richard Camplejohn, Daryll Green and Brian Shenton for reading the first draft of the handbook and for their helpful comments.

M.G. Ormerod

Safety

Attention to safety aspects is an integral part of all laboratory procedures, and both the Health and Safety at Work Act and the COSHH regulations impose legal requirements on those persons planning or carrying out such procedures.

In this and other Handbooks every effort has been made to ensure that the recipes, formulae and practical procedures are accurate and safe. However, it remains the responsibility of the reader to ensure that the procedures which are followed are carried out in a safe manner and that all necessary COSHH requirements have been looked up and implemented. Any specific safety instructions relating to items of laboratory equipment must also be followed.

Contents

7. Other Applications 53

Appendices 65

Index 75

Abbreviations

7-AAD	7-aminoactinomycin D
ADC	analogue to digital converter
AO	acridine orange
BrdU	5'-bromodeoxyuridine
cv	coefficient of variation
Cy-chrome	phycoerythrin–cyanine 5
DAPI	4', 6-diamidino-2-phenylindole
DCFH	2', 7'-dichlorofluorescin
EB	ethidium bromide
ECD	phycoerythrin–Texas Red
FDA	fluorescein diacetate
FISH	fluorescence *in situ* hybridization
FITC	fluorescein isothiocyanate
LI	labelling index
LWP	long wavelength pass
MClB	monochlorobimane
MESF	molecules of equivalent soluble fluorochrome
PBS	phosphate-buffered saline
PE	phycoerythrin
PerCP	peridinin–chlorophyll
PI	propidium iodide
plm	percent labelled mitoses
SD	standard deviation
SWP	short wavelength pass
TdT	terminal deoxynucleotidyl transferase

1 Introduction

1.1 Why flow cytometry?

In the last 10 years, the applications of flow cytometry have spread through all branches of biological sciences. Most research institutes now have several machines and even quite small research groups expect to have access to an instrument. People have used flow cytometers to measure the properties of and to sort mammalian and plant cells, yeast and bacteria and isolated nuclei, chromosomes and mitochondria. The biggest growth has been in the clinical field. Flow cytometers are now being used for routine measurements in immunology, haematology and, to a lesser extent, pathology departments.

In cells, a variety of properties may be measured, for example the DNA content of a nucleus, the expression of a surface antigen, the activity of an intracellular enzyme or the pH. Several properties might be measured simultaneously. Indeed, the scope of the technique is only limited by the fluorescent dyes available and the investigator's imagination – the latter probably being the most important.

1.2. What is flow cytometry?

As its name implies, flow cytometry is the measurement of cells in a flow system which has been designed to deliver particles in single file past a point of measurement. Although, in theory, many types of measurement could be made, in practice the term is applied to instruments which focus light on to cells and record their fluorescence and the light scattered by them. Electronic cell volume and absorbed light additionally may be measured.

The power of flow cytometry lies in the ability to measure several parameters on tens of thousands of individual cells within a few minutes. The method can therefore be used to define and to enumerate accurately sub-populations. Once identified, such sub-populations can be sorted physically for further study.

Typically, five parameters might be measured on 20 000 cells. Using blue light for excitation, one might record green, orange and red fluorescence and blue light scattered in a forward direction and at right angles to the laser beam. The large amount of data generated cannot be processed adequately without a powerful, well-programmed computer. The computer is not an optional extra but an essential part of the instrument.

The major disadvantage of flow cytometry is that a preparation of single particles (cells, nuclei, chromosomes) is required. Of necessity, tissue architecture is destroyed, so that spatial information about the relationships of cells to each other is lost. Also, no information is acquired about the distribution of entities within a cell and little, if any, information about a cell's shape. Flow cytometry can be contrasted to conventional light microscopy. The eye makes qualitative estimates of large numbers of parameters on a few cells, recording detail within each cell; the cytometer quantifies an average parameter for each cell but measures thousands of cells.

A comparison may also be made with a biochemical measurement in which an average value of, for example, an enzyme activity, is made for all the cells in a sample. Flow cytometry makes a measurement on each cell individually so that, if a small sub-set of cells has a high value, this feature will be recorded. Such a sub-set would not be detected biochemically.

Several recent books have given overviews of flow cytometry (Bauer *et al.*, 1993; Givan, 1992; Melamed *et al.*, 1990; Shapiro, 1988; Watson, 1991) and others have given detailed descriptions of many of the methods in common use (Darzynkiewicz and Crissman, 1990; Macey, 1994; Ormerod, 1994; Radbruch, 1992; Robinson, 1993).

Further reading

Bauer KD, Duque RE, Shankey TV. (1992) *Clinical Flow Cytometry. Principles and Applications.* Williams and Wilkins, Baltimore.

Darzynkiewicz Z, Crissman HA. (eds) (1990) *Flow Cytometry. Methods in Cell Biology*, Vol. 33. Academic Press, San Diego.

Givan AL. (1992) *Flow Cytometry. First Principles.* Wiley-Liss, New York.

Macey MG. (ed.) (1994) *Flow Cytometry: Clinical Applications.* Blackwell Scientific, Oxford.

Melamed MR, Lindmo T, Mendelsohn MI. (eds) (1990) *Flow Cytometry and Sorting*, 2nd edn. Wiley-Liss, New York.

Ormerod MG. (ed.) (1994) *Flow Cytometry. A Practical Approach*, 2nd edn. IRL Press, Oxford.

Radbruch A. (ed.) (1992) *Flow Cytometry and Cell Sorting.* Springer-Verlag, Berlin.

Robinson JP. (ed.) (1993) *Handbook of Flow Cytometry Methods.* Wiley-Liss, New York.

Shapiro HM. (1988) *Practical Flow Cytometry*, 2nd edn. Alan R. Liss, New York.

Watson JV. (1991) *Introduction to Flow Cytometry.* Cambridge University Press, Cambridge.

2 Instrumentation

2.1 Introduction

A basic flow cytometer consists of a source of light, a flow cell, optical components to focus light of different colours on to the detectors, electronics to amplify and process the resulting signals and a computer (*Figure 2.1*).

There are two basic types of flow cytometer – those that only analyse cells and those that analyse cells and can also physically sort them. The former are simpler, cheaper and are usually supplied as a bench-top machine.

2.2 The flow cell

The flow cell lies at the heart of the instrument. Its purpose is to deliver the cells singly to a specific point at which the source of light is focused. This is achieved by injection of the sample into the centre of a stream of

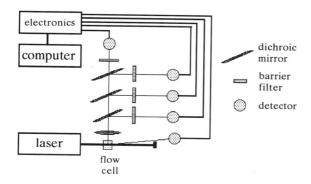

Figure 2.1: Layout of a typical flow cytometer. The detector in the forward direction would measure forward-scattered light; those collecting light orthogonally would measure four different colours, typically scattered primary light and green, orange and red fluorescences.

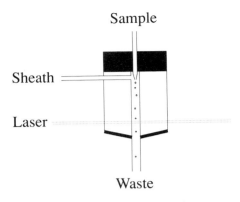

Sample

Sheath

Laser

Waste

Figure 2.2:   A typical flow cell.

liquid called the sheath fluid (normally either water or saline solution). The cell is designed so that the sheath fluid hydrodynamically focuses the sample stream delivering the cells to the point of detection with an accuracy of ± 1 µm or better.

There are three types of flow cell in general use. One ('cuvette') employs a quartz chamber (usually rectangular) placed at right angles to the laser beam with a channel cross-section of about 250 µm (*Figure 2.2*). This has the advantage of minimizing scattered light from the flow chamber. Fluorescence and light scatter can be measured over a wide angle at right angles to the laser beam and light scatter can also be measured in a forward direction with the addition of an obscuration bar to block the main laser beam. A typical speed of a cell through a flow chamber is 1 m sec^{-1}.

An alternative flow chamber is based on a microscope; in its simplest form, the sheath fluid and sample are squirted across a microscope slide. In a more sophisticated system, a channel is cut in a block replacing the microscope stage, the top surface of the chamber being a glass coverslip. Epi-illumination is used as in a conventional fluorescent microscope, that is, fluorescence is measured along the same optical path as the exciting light. It is more difficult to measure light scatter in this type of chamber.

The third design employs 'stream-in-air'. The sheath fluid, containing the hydrodynamically focused sample stream, emerges into the air from a nozzle just below which the laser beam is focused. The cells speed up as they emerge into air and travel at about 10 m sec^{-1}. Fluorescence and light scatter are measured as with the quartz chamber except that, because of the additional light scattered by the stream, a second obscuration bar has to be placed in the light collection path orthogonal to the laser beam. This design is generally used in cell sorters.

In a 'stream-in-air' configuration, the exit nozzle has an internal diameter of between 50 and 150 µm, 75 µm being typical. If a cuvette system is used in a cell sorter, the flow cell will also have a narrow exit nozzle. In analysers, a wider orifice can be used, avoiding problems caused by the blockage of the exit nozzle by larger particles (e.g. clumps of cells).

2.3 Light sources

The light source is normally either a laser or an arc lamp. Lasers are generally preferred because they produce monochromatic light and have a small 'spot' size. The latter is important because the light needs to be focused into a small volume to obtain maximum excitation of a single cell and also to ensure that, for most of the time, at an acceptable flow rate (approximately 1000 cells sec^{-1}), only one cell is in the laser beam.

Lasers filled with argon (argon-ion lasers) are found in nearly all laser-based instruments. A major line at 488 nm gives a convenient source of blue light for excitation of fluorescein, phycoerythrin and the newer tandem-conjugates for immunofluorescence and of propidium iodide (PI) for measurement of DNA (see Chapter 3, *Tables 3.1* and *3.2*). Air-cooled lasers producing 15 mW are now a common choice for bench-top cytometers. If other wavelengths are required (UV at 360 nm or, for example, 458 nm for excitation of chromosomes stained with chromomycin A_3), a more powerful, and expensive, water-cooled laser must be used. For excitation with UV alone, air-cooled helium–cadmium lasers producing 10 mW at 325 nm are available. Coherent Ltd have recently produced a laser containing a mix of argon and krypton which can be tuned to a wide range of wavelengths and which can be used in all line modes (i.e. without wavelength selection) so that two wavelengths can be used together.

Mercury arc lamps are used in some instruments. They are useful as an inexpensive source of UV light but do not give the sensitivity for the observation of weak immunofluorescence. With arc lamps, the correct wavelength of excitation must be selected using optical filters.

2.4 Optics

2.4.1 Focusing the excitation beam

The light beam must be focused on to the sample stream. This can be accomplished by a simple lens giving a beam cross-section of, typically, about 50 μm. Some instruments use a spherical cylindrical lens to produce a 20×60 μm elliptical beam. An alternative configuration is a crossed cylindrical pair of lenses which can produce an elliptical spot of, typically, 5×120 μm from a laser beam of 1 mm diameter.

Figure 2.3 shows a typical, Gaussian, profile from a laser beam focused with a spherical lens. If a cell passes through the beam slightly off-axis, it will experience a lower level of illumination and hence give less fluorescence. In an elliptically shaped beam, which has a flatter profile, this effect is minimized – an advantage for DNA analysis in which the operator

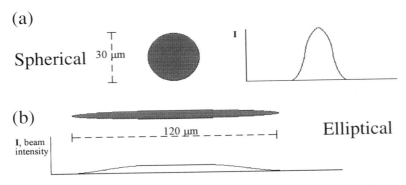

Figure 2.3:   Beam profiles of (a) spherically and (b) elliptically focused laser beams.

strives to eliminate any variation between cells caused by the instrument. (Other advantages of using an elliptically shaped beam for DNA analysis are discussed in Chapter 5.)

Spherical, or near-spherical, beams are used with stream-in-air systems in which the diameter of the beam has to be less than that of the stream in order to minimize excessive light scatter from the stream–air interface. The higher speed of the cells in this system gives an acceptably fast signal pulse (2–7 µsec) with the wide laser beam. The whole cell is illuminated as it passes through the beam and the loss of fluorescence due to the presence of an obscuration bar is compensated for partly by the more efficient excitation of the cells.

Wide, flat beams are used with flow chambers based on quartz chambers, with which more efficient collection optics can be used to compensate for the lower level of illumination of the cells. The small beam height also gives fast electronic pulses (approximately 5 µsec) from the more slowly moving cells.

2.4.2 Light collection

The orthogonal collection lens should have a high numerical aperture in order to collect as much of the fluorescence as possible. In analysers, as opposed to cell sorters, it is possible to use a shorter working distance, including immersion objectives, and hence a higher numerical aperture. In some instruments using a flow cell, a spherical mirror is mounted on the opposite side of the flow cell to the collection lens. The focal point of the mirror is the intersection of the sample stream and laser beam so that the amount of light collected is doubled.

For collection of forward scattered light, sensitivity is not a problem and a simple long-working-distance lens is satisfactory. Forward scatter is sensitive to the angle over which the scattered light is collected and will, therefore, depend on the geometry of the light collection. Consequently, scatter profiles may differ between instruments.

2.4.3 Optical filters

If the source of exciting light is an arc lamp, optical filters are needed to select the correct wavelength of excitation. These are usually made from coloured glass. Lasers give monochromatic light which does not need further filtration.

The dichroic and bandpass filters used on the output side are normally based on interference filters. The manufacturers' catalogues from Oriel and Melles Griot (see Appendix B) give excellent descriptions of their construction.

Bandpass filters transmit light over a narrow band and generally are used immediately in front of the detector. Their important parameters are the peak wavelength of transmission, the percentage of light transmitted at the peak wavelength and their bandwidth (measured as the separation of the 50% transmission points).

Dichroic filters (sometimes called beam splitters) are edge filters which are used in the flow cytometer at an angle of 45°. Short wavelength pass (SWP) filters transmit light below a given wavelength and reflect light of longer wavelengths. Long wavelength pass (LWP) filters work in the reverse fashion. Their important parameters are the wavelength for 50% transmission (the cut-off for LWP or the cut-on wavelength for SWP), the peak transmission and the slope at the cut-on or cut-off wavelength.

Filters are supplied with a graph of their transmission against wavelength. This measurement should have been made at the angle at which the filter is to be used. Commercial flow cytometers are fitted with sets of filters by the manufacturer. It is important that the users should be aware of their optical properties if they are to have an understanding of how spectral overlaps will affect measurements made in their instrument (see Section 3.3).

Table 2.1 gives the properties of filters and mirrors that might be used in the typical layout shown in *Figure 2.1*. This configuration would be suitable for measuring scattered light and green (fluorescein), orange (phycoerythrin) and red (peridinin–chlorophyll conjugate) fluorescences.

Table 2.1:   Three colour immunofluorescence excited by an argon-ion laser (488 nm). Typical arrangement of dichroic mirrors and bandpass filters.

Filter	Wavelength (nm)	Detection
First dichroic (long pass)	500	Select < 500 nm (scattered light)
Second dichroic (long pass)	560	Select 500–560 nm (fluorescein)
Third dichroic (long pass)	600	Select 560–600 nm (phycoerythrin)
First bandpass	488/10	Scattered light
Second bandpass	530/30	Fluorescein
Third bandpass	580/30	Phycoerythrin
Long pass filter	610	Peridinin–chlorophyll

A long pass dichroic mirror reflects light below the given wavelength passing light of longer wavelength.
The numbers for the bandpass filters give the wavelength of transmission/the 50% bandwidth.

A list of manufacturers of suitable filters is given in Appendix B. Further discussion of optical filters can be found in Ploem and Tanke (1987).

2.5 Detectors

A solid state detector is sufficient for the measurement of forward scattered light. For measuring fluorescence and orthogonal scatter, photomultipliers are used.

2.6 Signal processing

2.6.1 Electronic trigger

After pre-amplification, the signal from a photomultiplier undergoes further processing. The instrument must be set to respond to signals derived from the particle of interest (e.g. a cell) and to ignore debris and 'spikes' from electronic noise. A threshold level is set on one or, possibly, two parameters such that the electronics is only triggered if the signal rises above this level. It is usual to use light scatter for the trigger. An instrument may be triggered on a signal from a fluorescent stain for DNA but immunofluorescence generally should never be used since negative cells might inadvertently be excluded from analysis.

2.6.2 Pulse shape analysis

If the width of the laser beam is greater than that of the cell diameter, the peak of the signal is usually recorded. If a narrow elliptically shaped beam is used, only part of a cell may be illuminated at any given time. To produce a signal proportional to the total fluorescence of the cell, the pulse is integrated (pulse area). The width and peak of the pulse may also be recorded and this will give some information about the length of the cell passing through the beam. This information may be used in a DNA measurement to distinguish between single cells or nuclei and doublets (see Chapter 5; also Ormerod, 1994; Watson, 1991).

2.6.3 Amplification

Most cytometers offer a choice between linear and logarithmic amplifiers. For DNA measurement, linear amplification should always be used. For immunofluorescence, logarithmic amplification increases the dynamic range so that weak and strong signals can be recorded on the same scale. (For further discussion, see Section 2.7.3 below.)

The linearity of the response of the amplifiers should be checked regularly using standard beads. A DNA histogram gives a simple linearity check since the channel of the peak from cells in G2 should be exactly double that of the cells in G1. A non-linear response might be caused by non-linearity of the amplifier or by a non-zero offset on the amplifier. In instruments using 'stream-in-air', a badly adjusted blocker bar can be a cause of non-linearity.

2.6.4 Analogue to digital conversion

The signals undergo analogue to digital conversion before being transmitted to the computer. The resolution of the data will depend on the analogue to digital converter (ADC) – an 8-bit converter will yield 256 channels; a 10-bit converter gives 1024. Most manufacturers have now moved over to using the latter.

2.7 Data analysis

2.7.1 Methods for displaying data

The different parameters are displayed on the computer screen as either univariate (*Figure 2.4a*) or bivariate histograms (cytograms). The latter can be shown in various ways. If data are being displayed in real time, they are shown as a 'dot plot' (*Figure 2.4b*). Subsequently they may be drawn as a contour plots (*Figure 2.4c*) or as pseudo-three-dimensional isometric plots (*Figure 2.4d*). The univariate histograms and cytograms are stored on computer disc.

As well as being displayed in real time, raw data can be written in a continuous stream on to a disc. This so-called listed data can subsequently be re-analysed in more detail.

2.7.2 Gating

Most modern instruments are capable of analysing at least five parameters. Clearly, all the parameters cannot be displayed in a correlated fashion. To make full use of the information collected, 'gating' is employed. Data from one or two parameters are displayed, regions of interest (gates) are defined to select certain populations of cells for display of further parameters.

An example is shown in *Figure 2.5*, which shows data recorded from nuclei extracted from a paraffin-embedded block of tissue from a human breast carcinoma. The nuclei were incubated with PI, which stains DNA.

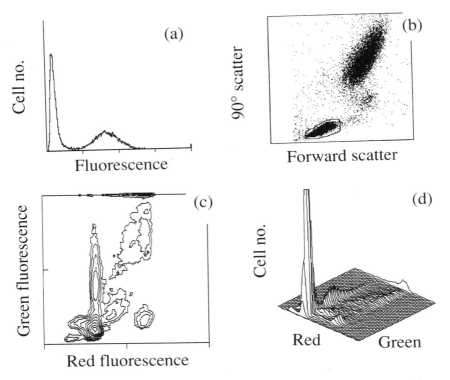

Figure 2.4: Typical displays from a flow cytometer. (a) A single parameter histogram showing negative and positive populations from rat lymphocytes incubated with a fluorescein isothiocyanate (FITC)-labelled antibody specific for thymocytes. (b) A 'dot plot' (or cytogram) showing 90° vs. forward angle light scatter for human peripheral blood leukocytes; a region has been set on the cluster defining lymphocytes (Becton-Dickinson FACScan; data recorded during the Royal Microscopical Society course in Cambridge). (c) A contour plot of a cytogram of green (FITC) vs. red (DNA/PI) fluorescence for immature rat thymocytes labelled 8 h previously with 5'-bromodeoxyuridine (BrdU) (see Section 6.2, for further details) (data recorded on an Ortho Cytofluorograph, excitation at 488 nm; cells prepared by Howard Fernhead, MRC Toxicology Unit, Leicester). (d) The data shown in (c) plotted as a pseudo-three-dimensional plot.

Forward and orthogonal (90° or side) light scatter were recorded together with red (PI) fluorescence. The cytogram of light scatter clearly identified two populations of nuclei. Regions were set to define the sub-populations and their DNA (red PI fluorescence) histograms recorded separately. The nuclei scattering more light at 90° were aneuploid (from tumour cells) while those scattering less were diploid (from normal cells). From the ungated DNA histogram, it would be concluded that the tumour was diploid with either a high G2M phase or a tetraploid component. However, gating on light scatter has revealed that the tumour is hypoploid with a DNA content 7% below normal. The nuclei of high DNA content have also been resolved.

The advantage of a computer which can carry out this type of analysis in real time cannot be over-emphasized. The cytograms or histograms of

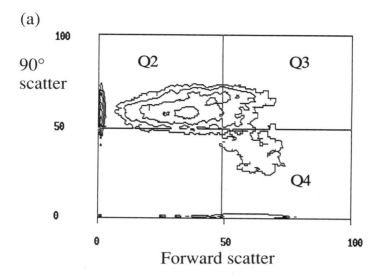

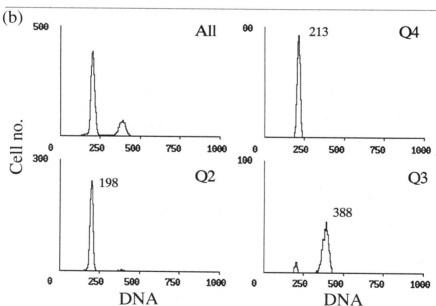

Figure 2.5: An illustration of gating. Nuclei extracted from a formalin-fixed, paraffin-embedded human breast tumour and stained with PI. The cytogram (a) shows the light scatter. Quadrant regions were set as shown (b) and the DNA histograms for all the nuclei and for nuclei falling within three of the quadrants are shown. Nuclei with low 90° scatter in quadrant 4 were from normal cells; those with high 90° scatter (quadrants 2 and 3) were from tumour cells. The channel numbers of the major peaks are shown for the gated data. There were both hypo- and hyperdiploid nuclei present in the tumour. (Data recorded on an Ortho Cytofluorograph, excitation at 488 nm and the nuclei prepared by Mrs J. Titley, Institute of Cancer Research.)

interest need only be saved to disc taking up less storage space than listed data and avoiding having to spend time after the experiment in further analysis. Any unusual features of the data can also be recognized more quickly.

2.7.3 Logarithmic vs. linear amplification

An example of the different displays generated by the methods of amplification are show in *Figure 2.6*. The logarithmic amplifier has the effect of expanding weak signals and compressing strong signals. If the amplifier gain is altered, the width of the signal (i.e. its standard deviation, SD) will be independent of the gain setting. With a linear amplifier, the SD will increase with increasing gain.

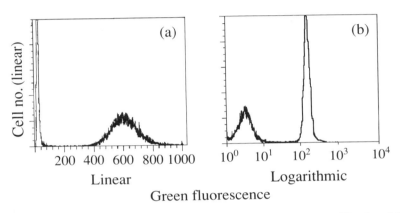

Figure 2.6: The difference between logarithmic and linear amplification. (a) Linear and (b) logarithmic amplifier. Figure drawn from data supplied by N. P. Carter.

2.7.4 Simple statistical analysis

The computer should give percentage of total count, mean and median channels and SD of the data defined in a region. If a logarithmic amplifier has been used, the data should be transformed to a linear form before calculating a mean or SD. A common measure of the spread of a distribution is the coefficient of variation (cv) given by:

$$cv = SD/mean\ channel \times 100\%.$$

This parameter is used because it is independent of channel number (dimensionless) and allows samples from different instruments on different settings to be compared directly.

2.8 Cell sorting

The commonest method of sorting cells is by electrostatic deflection of charged droplets. A conductive sheath fluid is used (buffered saline). The flow cell is vibrated vertically by means of a piezoelectric transducer at a frequency in the order of 30 kHz. This vibration causes the fluid emerging from the exit nozzle (typically 75 μm diameter) to break up into droplets. The flow cell is charged (± v where v typically lies between 50 and 150 V) at the moment a cell of interest is inside the droplet currently being formed. The stream of droplets passes through a pair of charged plates (5000 V) so that droplets which are charged are deflected and collected together with the cell contained therein (*Figure 2.7*).

To ensure that the flow cell is charged at the correct moment, the time delay between a cell passing through the laser beam and the droplet break-off point has to be determined. Anything that influences the position

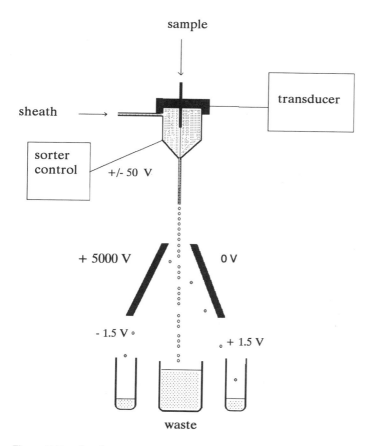

Figure 2.7:   A cell sorter.

of the break-off point (a change in temperature, a draught, dirt in the flow cell orifice) will adversely affect the stability of the sorter.

Deflecting and collecting more than one droplet for each cell sorted will minimize the influence of small changes in sorting conditions. Occasionally there will be more than one cell in the deflected droplets. The electronic circuitry can detect these coincidences and, if high purity (at the expense of a slightly lower yield) is desired, the sort decision can be aborted.

Under ideal conditions, cells can be sorted with a purity of 98% or better. Although the purity is high, the yield is low compared to many other methods of cell separation. Assuming a flow rate of 1000 cells sec^{-1}, if the concentration of the sub-population to be sorted is 10% and allowing for losses through rejection of coincidences, the maximum number of cells collected will be about 3×10^5 h^{-1}.

A different approach to cell sorting is employed in the Partec PAS III cell sorter and the Becton-Dickinson FACSort. In the PAS III, a piezoelectric fluidic valve operates on one arm of a Y-shaped flow channel, the cells to be collected being deflected down one arm of the Y. The sorter can be used for sorting large particles, such as protoplasts, because it avoids the limitations of size imposed by the need for accurate droplet formation in the electrostatic sorters. In the FACSort, a piezoelectric device deflects a small collector into the sample stream to gather the selected cell.

Further reading

Carter NP, Meyer EW. (1994) Introduction to the principles of flow cytometry. In *Flow Cytometry. A Practical Approach*, 2nd edn. (ed. MG Ormerod). IRL Press, Oxford, pp. 1–28.

Omerod MG. (ed.) (1994) *Flow Cytometry. A Practical Approach*, 2nd edn. IRL Press, Oxford.

Ploem JS, Tanke HJ. (1987) *Introduction to Fluorescence Microscopy*. Royal Microscopical Society Handbook no. 10. Oxford University Press, Oxford.

Van Dilla MA, Dean PN, Laerum OD, Melamed MR (eds) (1985) *Flow Cytometry Instrumentation and Data Analysis*. Academic Press, Orlando.

Watson JV. (1991) *Introduction to Flow Cytometry*. Cambridge University Press, Cambridge.

Wood JCS. (1993) Clinical flow cytometry instrumentation. In *Clinical Flow Cytometry. Principles and Applications* (eds KD Bauer, RE Duque, TV Shankey). Williams and Wilkins, Baltimore, pp. 71–92.

3 Fluorescence

3.1 Introduction

Fluorescence occurs when a molecule excited by light of one wavelength loses its energy by emitting light of a longer wavelength. Because the colours of the exciting and emitting light are different, they can be separated from one another by using optical filters.

Fluorescence detection is a sensitive technique because a positive signal is observed against a negative background. Fluorescence and light scatter are the two types of measurement made in most flow cytometers. The detection of at least three compounds fluorescing at different wavelengths permits multiparametric analysis of cells and has greatly increased the power of flow cytometry.

3.2 Properties of a fluorophore

The important properties of a fluorophore are its absorption spectrum, its extinction coefficient at a wavelength convenient for excitation (e.g. the blue line from an argon-ion laser at 488 nm), its emission spectrum and its quantum efficiency. The latter is the number of photons emitted for every photon absorbed. The difference between the maxima in the wavelengths of absorption and emission is known as the Stokes' shift.

The properties of a fluorophore will depend on its environment. For example, PI which is used to stain DNA is only weakly fluorescent in water; on intercalating with DNA, the fluorescence increases 50-fold due to the hydrophobic environment. Some fluorophores, such as fluorescein, are sensitive to pH. There are other compounds whose fluorescent spectrum changes according to the concentration of Ca^{2+} ions (see Section 3.4.3).

If two fluorophores are closely associated, energy transfer can occur, whereby excitation of one compound causes the other to fluoresce. For example, if the rhodamine derivative, Texas Red, is bound to the fluorescent protein, phycoerythrin (PE), excitation of PE will lead to emission of light from Texas Red. The effective emission maximum will have been shifted from that of PE (576 nm) to that of Texas Red (620 nm).

Fluorescence can also be quenched by interaction with another molecule, the energy being dissipated by a non-radiative transition. This can be illustrated by the bis-benzimadazole dye, Hoechst 33342, which binds to DNA, giving blue fluorescence on excitation with UV. However, if cells are labelled with BrdU, the fluorescence of the dye is quenched by the bromine atom (see Chapter 6). If a compound is over-labelled with a fluorophore, fluorescence can also be quenched by interactions between the molecules of fluorophore. This effect can be observed when liposomes are loaded with fluorescein; fluorescence decreases with increasing dye concentration above about 20 μM.

3.3 Spectral overlap

Fluorochromes have a wide emission spectrum. When multiple fluorescences are measured, there will inevitably be spectral overlap between them. For example, using the combinations shown in *Table 2.1*, some of the fluorescein emission will be collected by the PE detector. *Figure 3.1* shows the emission spectra for two commonly used antibody labels, fluorescein and PE. Superimposed on them are the possible transmission characteristics of two barrier filters chosen to give high sensitivity. The fluorescein emission spectrum overlaps with that of PE and some of its light will be transmitted by the PE filter. This spectral overlap is usually

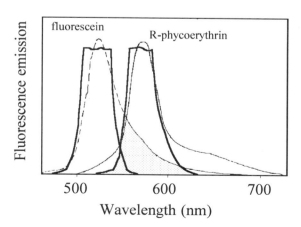

Figure 3.1: Demonstration of the spectral overlap between fluorescein and phycoerythrin. The thin lines are the emission spectra of fluorescein and phycoerythrin; excitation wavelength: 488 nm. The thick lines show the characteristics of two barrier filters. The shaded area represents light from fluorescein which will be collected by the PE filter.

corrected by subtracting a fraction of the fluorescein signal from the PE signal and vice versa.

In most published data, a correction has been applied for this spectral overlap either electronically or in the computer software.

3.4 Fluorophores used in flow cytometry

3.4.1 Compounds used to label proteins covalently

There are two classes of fluorophore used in flow cytometry – those which are covalently bound to other probes (almost invariably a protein) and those which bind non-covalently to structures within the cell. The largest application for labelled proteins is for immunofluorescence. Monoclonal antibodies may be labelled directly or indirectly. The latter is achieved either with a labelled antibody to immunoglobulin or by labelling the primary antibody with biotin and using fluorescently labelled streptavidin which has four binding sites for biotin (methods of labelling antibodies are illustrated by Polak and Van Noorden, 1984; Ploem and Tanke, 1987).

Some of the more widely used fluorescent labels are listed in *Table 3.1*. The most common fluorophore used to label proteins is fluorescein iso-thiocyanate (FITC). Labelling is achieved by reaction of the isothiocyanate with lysine residues on proteins; a similar strategy is adopted for Texas Red. PE and allophycocyanin are large proteins and a different method has to be used. They can be joined directly to immunoglobulins with a linker molecule such as *N*-succinimidyl 3-(2-pyridyldithio) propionate. Alternatively they can be labelled with biotin and combined with a biotin-labelled antibody with either avidin or streptavidin as a bridge.

A wide range of directly labelled monoclonal antibodies and labelled anti-immunoglobulins are available from several manufacturers. It is seldom necessary to label proteins in the laboratory and the reader is recommended to purchase labelled reagents.

Table 3.1: Properties of some fluorophores used to label proteins

Fluorophore	Excitation maxima (nm)	Emission maximum (nm)
Fluorescein	495	520
R-phycoerythrin	495, 564	576
Texas Red	596	620
Phycoerythrin–Texas Red conjugate	495	620
Phycoerythrin–cyanine5 conjugate	495	670
Conjugated peridinin chlorophyll	490	677
Allophycocyanin	650	660
Coumarin	357	460

It should be noted that the wavelengths of absorption and emission may depend on the nature of the conjugate used and its environment.

3.4.2 **Probes for nucleic acids** (*Table 3.2*)

The fluorophores described in this section bind stoichiometrically to nucleic acids so that they can be used for quantitative measurement. This is essential for the measurement of ploidy and the cell cycle (see Chapter 4).

The bis-benzimadazoles, such as Hoechst 33342, cross an intact plasma membrane and can be used to visualize the cell cycle in viable cells. Cells have to be fixed or permeabilized before staining for DNA with the other compounds listed.

Propidium iodide (and the closely related compound, ethidium bromide) intercalates between the bases in double-stranded nucleic acids. It is excited by blue light, giving red fluorescence. With an argon-ion laser tuned to 488 nm, PI can be used in combination with fluorescein which makes it particularly useful for simultaneous measurement of antibody binding and DNA content. PI also binds to double-stranded RNA which has to be removed by treatment with RNase.

Acridine orange fluoresces green when intercalated in double-stranded nucleic acids and red when stacked on the charged phosphates in single-stranded nucleic acids. These properties permit the simultaneous measurement of DNA (double-stranded) and RNA (single-stranded) (Darzynkiewicz and Kapuscinski, 1990). A related compound, pyronin Y, is used to label RNA after its binding sites to DNA have been blocked by a non-fluorescent dye or the DNA removed by treatment with DNase.

The bis-benzimadazoles (usually referred to by the number given to them by the manufacturer, Hoechst) bind preferentially to AT-rich regions in the small groove of double-stranded DNA and fluoresce blue when excited by UV light. The antibiotics, mithramycin and chromomycin A_3, bind to the GC-rich regions of DNA. Chromomycin has been used in conjunction with Hoechst 33258 to stain chromosomes in order to resolve chromosomes of similar size but with different AT/GC ratios (see Chapter 7).

4′,6-Diamidino-2-phenylindole (DAPI) has been used extensively to label DNA for excitation with UV light; it is often the stain of choice in instruments employing a mercury arc lamp.

Table 3.2: The properties of some fluorophores used to label nucleic acids

Fluorophore	Excitation maxima (nm)	Emission maximum (nm)
Propidium iodide	495, 342	639
Ethidium bromide	493, 320	637
Acridine orange	503	530 (DNA)
		640 (RNA)
Mithramycin	445	569
Chromomycin A_3	430	580
Hoechst 33342	395	450
DAPI	372	456
Pyronin Y	545	565

The fluorescent properties of most of these dyes change on binding to nucleic acid. The wavelengths given are those of the dye–nucleic acid complex.

3.4.3 Other reporter molecules

Applications using other reporter molecules are outlined in Chapter 7. A few general principles are described below.

Loading the cell.   The internal milieu of the cell can be explored using compounds whose fluorescence is sensitive to the environment. Such reporter molecules must be introduced into the cell and, once there, they should not migrate out. This is not a problem if the probe binds tightly to a cell component (e.g. Hoechst 33342 to DNA) or if the partitioning of the probe between the cytoplasm and the medium is being measured (e.g. probes of plasma membrane potential, see below). With other probes a different strategy must be found.

Many reporter molecules are charged and, as such, have difficulty in crossing the plasma membrane; a general procedure adopted for loading cells is to mask the charge by preparing an ester. The uncharged molecule will diffuse into the cell in which non-specific esterases convert the reporter molecule to its original non-permeant form (*Figure 3.2*).

Ratio measurements.   Once in the cell, the amount of a reporter molecule present, and hence its total fluorescence, will generally depend on the size of the cell. Changes in the amount of light emitted on a single cell basis cannot be measured; only a general shift in the population can be recorded. It is therefore useful to use a probe which shows a shift in the wavelength of emission. Fluorescence is measured at two wavelengths and the ratio calculated for each cell. This ratio will be independent of the total amount of fluorophore in the cell.

For example, the calcium indicator, indo-1, has an emission maximum of about 490 nm in the absence of Ca^{2+}; in the presence of 1 mM Ca^{2+}, this shifts down to about 410 nm (*Figure 3.3*). It can be loaded into cells as its acetoxymethyl ester (see above). Indo-1 is excited with UV light and

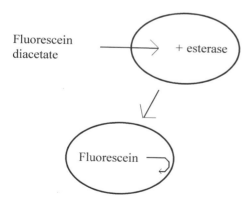

Figure 3.2:   Intracellular conversion of fluorescein diacetate to fluorescein.

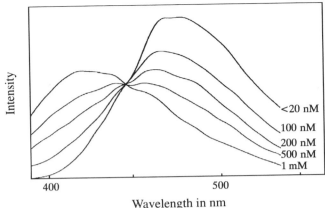

Figure 3.3: The change in the emission spectrum of indo-1 with changes in Ca^{2+} concentration; excitation wavelength, 360 nm. Redrawn from Rabinovitch and June (1994) with permission from Oxford University Press.

changes in the concentration of intracellular Ca^{2+} recorded by observing fluorescence at 400 and 520 nm.

Reporter molecules showing a shift in emission spectrum have also been used to measure intracellular magnesium ions and pH (see Section 7.10).

References

Darzynkiewicz Z, Kapuscinski J. (1990) Acridine orange: a versatile probe of nucleic acids and other cell constitutents. In *Flow Cytometry and Sorting*, 2nd edn. (eds MR Melamed, T Lindmo, MI Mendelsohn). Wiley-Liss, New York, pp. 291–314.

Ploem JS, Tanke HJ. (1987) *Introduction to Fluorescence Microscopy*. Royal Microscopical Society Handbook no. 10. Oxford University Press, Oxford.

Polak JM, Van Noorden S. (1984) *An Introduction to Imunocytochemistry: Current Techniques and Problems*. Royal Microscopical Society Handbook no. 11. Oxford University Press, Oxford.

Rabinovitch PS, June CH. (1994) Intracellular ionised calcium, magnesium, membrane potential and, pH. In *Flow Cytometry. A Practical Approach*, 2nd edn. (ed. MG Ormerod). IRL Press, Oxford, pp. 185–215.

Further reading

Ormerod MG. (1994) An introduction to fluorescence technology. In *Flow Cytometry. A Practical Approach*, 2nd edn. (ed. MG Ormerod). IRL Press, Oxford, pp. 217–413.

Waggoner AS. (1990) Fluorescent probes for cytometry. In *Flow Cytometry and Sorting*, 2nd edn. (eds MR Melamed, T Lindmo, MI Mendelsohn). Wiley-Liss, New York, pp. 209–225.

The catalogue from Molecular Probes, Inc., is an excellent source of information about most fluorescent probes used in flow cytometry.

4 Immunofluorescence

4.1 Introduction

The most common routine application of flow cytometry is the measurement of surface antigens by immunofluorescence labelling using monoclonal antibodies. Bench-top flow cytometers are designed for this application and are finding increasing use in clinical laboratories for enumerating lymphocyte sub-sets from peripheral blood and for typing leukaemias and lymphomas in samples of blood and bone marrow.

4.2 Fluorochromes used for immunofluorescence

The most popular label for antibodies is fluorescein, which has an excitation maximum at 495 nm and emits green light (520 nm). It can be excited conveniently by an argon-ion laser tuned to 488 nm. The power of an analysis can be extended by measuring two, three or even four immunofluorescences simultaneously with a single argon-ion laser.

For multicolour analysis, the second label is normally R-phycoerythrin, a phycobiloprotein extracted from red algae (excitation maxima, 564 and 495 nm; emission maximum, 576 nm). A third antigen can be visualized using one of the several fluorochromes available which fluoresce red when excited at 488 nm, such as phycoerythrin–Texas Red (ECD), phycoerythrin–cyanine5 (Cy-chrome) or peridinin–chlorophyll (PerCP) conjugates (see *Table 3.1*). The fluorescences from ECD and either Cy-chrome or PerCP are sufficiently separated to allow them to be used together, thus permitting four-colour analysis. As the range of available dyes increases, one laser, four-colour analysis will become increasingly common.

It is important that the cells of interest are selected by gating on a cytogram of side vs. forward scatter. In a study of cultured cells, the gate may just exclude clumps, debris and possibly dead cells, which often have lower forward scatter. In multicellular systems, a particular type of cell may be delineated – for example, in peripheral blood, lymphocytes can be selected to the exclusion of monocytes and granulocytes (*Figure 4.1*). If there is a problem in identifying the lymphocyte cluster, then a process called 'back gating' can be used. The lymphocytes are labelled specifically with an appropriate antibody, a gate is set on a histogram of fluorescence and a cytogram of the light scatter of the positive cells displayed (*Figure 4.1f*). A gate set on this cluster can now be used to delineate the

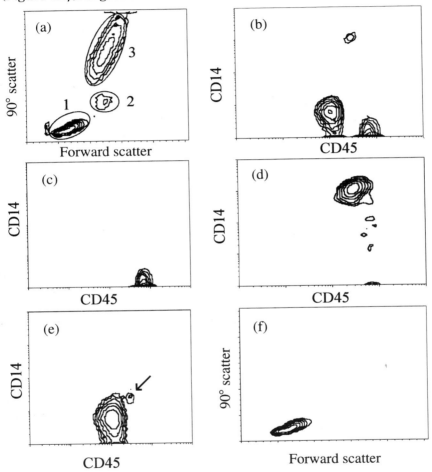

Figure 4.1: Peripheral blood leukocytes labelled with CD4–FITC and CD8–PE. Becton-Dickinson FACScan employing a 15 mW argon-ion laser tuned to 488 nm. (a) Cytogram of 90° vs. forward light scatter. Linear amplification. A gate has been set on the three major clusters defined by light scatter (1, lymphocytes; 2, monocytes; 3, granulocytes). (b) Cytogram of green (CD14) vs. orange (CD45) fluorescence of all the cells. Logarithmic amplification. (c) Cytogram of CD14 vs. CD45 fluorescence of the lymphocytes only (gated on light scatter, region 1). (d) Cytogram of CD14 vs. CD45 fluorescence of the monocytes only (region 2 of light scatter). (e) Cytogram of CD14 vs. CD45 fluorescence of the granulocytes only (gated on region 3). (f) 'Back gating'. Light scatter profile obtained by gating on the CD45^{bright}, CD14$^-$ cells. Data supplied by Dr Mark Lowdell, The Royal Free Hospital, London, and analysed using PCLYSIS software (Becton-Dickinson).

lymphocytes in a cytogram of light scatter from all the cells. A combination of antibodies to CD45 (leukocyte common antigen) and CD14 are often used for this purpose. Lymphocytes are CD45^{bright}, CD14$^-$, monocytes CD45$^+$, CD14$^+$, and granulocytes CD45^{dim}, CD14$^-$ (see *Figure 4.1*). The small additional cluster arrowed in *Figure 4.1e* probably arises from eosinophils whose light scatter is slightly greater than the granulocytes.

In blood and, particularly, bone marrow samples from diseased patients, care should be exercised in selecting the correct gate since the cells of pathological significance may show an abnormal pattern of light scatter. Back gating from a cluster of abnormal fluorescence may help but this procedure is not without pitfalls. The choice of gate can often be guided by a careful examination of a stained blood film by conventional light microscopy looking for atypical morphology.

Dead cells can be identified by adding a DNA-binding dye, such as PI, which only stains cells with a damaged plasma membrane. Overlap between the strong, red fluorescence from the dead cells and other dyes, such as PE, is unimportant since the dead cells are excluded from further analysis.

4.3 Sample preparation and staining

As with all flow cytometric measurements, sample preparation is of paramount importance. A suspension of live, single cells with little debris and few clumps is essential. Erythrocytes should be removed from clinical samples (peripheral blood, bone marrow); removal by differential lysis is preferred to separation by centrifugation on a Ficoll–sodium diatrizoate gradient because the latter is more expensive and time-consuming and varying numbers of cells may be lost during processing. Several companies (for example, Becton-Dickinson, Coulter and Ortho) supply reagents for red cell lysis as part of the washing procedure used after labelling with the antibodies. The reagents used for lysis of erythrocytes affect the leukocytes and the light scatter profile (*Figure 4.1a*) will depend on the method employed.

The majority of antibodies used for flow cytometry are murine mono-clonal immunoglobulins. Antibodies in general use, particularly those directed against the human cluster of differentiation groups (CD), are supplied either unlabelled, labelled with biotin or labelled directly with FITC or PE.

For single colour analysis, cells may be labelled using either direct or indirect staining, the methods being similar to those used for immunocyto-chemistry (Polak and van Noorden, 1984). The direct method has the advantage of speed; indirect methods give brighter fluorescence and the reagents are usually cheaper. When using an indirect method with a second antibody, it can be advantageous to use a labelled F(ab′)₂ fragments to prevent the second reagent binding to Fc receptors on, for example, monocytes.

For two-colour analysis, if the antibodies have been raised in different species, indirect labelling can be used after checking that the second antibodies show no cross-species reactions. If, as is common, both antibodies are murine monoclonal IgGs, either direct staining is used for both antibodies or one antibody is labelled directly with FITC and the other with biotin followed by PE–avidin. The correct colour compensation must be applied to adjust for spectral overlap between the two fluorochromes (see Section 3.3); be aware that over-correction can give misleading results.

Similar comments apply to three-colour analysis except that colour compensation has now to be applied to three channels (usually for green into orange, orange into green and orange into red). Some thought should be given to the selection of fluorochromes for the different antibodies being used in order to minimize any problem caused by spectral overlap and the differences in antigen distribution on the various types of cell. For example, a strong fluorescein signal will tend to swamp a weak signal from PE. Fluorescein would be the dye of choice for the weakest staining antibody in a dual combination.

After labelling, samples can be fixed to enable them to be stored for later analysis or to inactivate viruses (important in clinical samples, particularly those from HIV-positive patients). Cell pellets are normally fixed in 0.5–2% paraformaldehyde in phosphate-buffered saline (PBS) – conditions which preserve the light scattering properties of the cells. In the analysis of one- or two-colour fluorescences, cells which were damaged before fixation can be identified by the addition of the laser dye, LDS-751 (Exciton), which binds to the DNA of the damaged cells and fluoresces red (Terstappen *et al.*, 1988).

Nuclear or cytoplasmic antigens can also be labelled after the cells have been fixed or permeabilized. The optimum conditions of fixation need to be determined for each particular antigen. Fixatives commonly used include ethanol, methanol and paraformaldehyde; unfixed cells can be permeabilized using a variety of detergents (Clevenger and Shankey, 1993; Larsen, 1994). Faced with a new antibody, the following procedures might be tested (Camplejohn, 1994, and personal communication; see also Larsen, 1994; Schimenti and Jacobberger, 1992).

1. Fixation in 70% ethanol at 0°C.
2. Fixation in absolute methanol at – 20°C.
3. 1% paraformaldehyde followed by methanol, both at 0°C.
4. Incubation of fresh cells with antibody in the presence of detergent at 0°C.

and for nuclear antigens only:

5. A detergent enucleation technique followed by steps 2 or 3 above.

A suitable enucleation technique is given by Petersen (1985).

When the immunofluorescence is weak, it is sometimes difficult to determine whether there is positive staining above the negative background. Unlabelled cells will exhibit autofluorescence and, at high amplifier gains, 'break-through' of the primary light might make a contribution (particularly with a set of old filters). In an indirect staining method, the second antibody on its own might react non-specifically with the cells. The best control for non-specific staining is to use an immunoglobulin of the same isotype as the staining antibody. In a direct method, the control immunoglobulin has to be labelled with the appropriate fluorochrome.

4.4 Quantitative analysis

4.4.1 Counting cells

It can be important to know the absolute number of a cells in a particular sub-set. For example, the progression of AIDS is followed by enumerating CD4-positive lymphocytes in the peripheral blood. Flow cytometers measure accurately the relative percentage of cells in a sub-set but they are not designed to measure cell concentrations.

An accurately measured volume of a sample can be 'doped' with a known number of beads which can be distinguished easily on the flow cytometer in a cytogram of forward vs. right angle light scatter. Absolute cell counts can be obtained by measuring the relative percentages of the beads and the cellular sub-set of interest.

An alternative for use with samples from patients with AIDS is to measure the concentration of lymphocytes present with a haematology counter and then use the flow cytometer to measure the percentage of CD4-positive cells. This method has the disadvantage of compounding any intrinsic errors in the two machines.

In one flow cytometer (the Ortho Cytoron) the sample is injected with a syringe delivering a controlled volume. The instrument records the number of cells which have been examined and, from the relative number of CD4-positive lymphocytes, the absolute concentration can be calculated.

4.4.2 Measuring the number of antigenic sites on the cell surface

Measurement of the absolute number of antigens on the cell surface can be useful in the sub-classification of lymphomas and leukaemias and in defining pathological conditions, such as immune deficiencies. Quantification requires the conversion of fluorescence intensity into the number of antibody molecules bound to the cell. Fluorescence is dependent on the environment of the probe, particularly pH, and any conversion needs to accommodate environmental factors.

To compare the amount of a given antibody bound to different cells, the fluorescence can be quantified using a standard unit called molecules of equivalent soluble fluorochrome (MESF). A measurement of MESF gives the intensity of the signal from the sample relative to a solution of fluorochrome molecules. A calibration can be established using pre-calibrated beads purchased from Flow Cytometry Standards Corporation and the MESF of the sample read from the curve.

Beads carrying anti-mouse IgG can also be obtained from Flow Cytometry Standards Corporation. The beads have been calibrated so that, at saturation, they will absorb known amounts of antibody and hence, in conjunction with a measurement of MESF, can be used for quantification in both direct and indirect techniques.

For indirect labelling using murine IgG monoclonal antibodies, Biocytex sell beads carrying measured amounts of murine IgG accessible to a secondary reagent. The secondary reagent (which can be labelled with any fluorochrome) is used on the beads at the same concentration under the same conditions as used for the cells, and a calibration curve of fluorescence intensity vs. antibody density is established. Again the primary antibody must be used in saturating quantities.

4.5 Quality control

The general performance of the instrument should be checked daily using standard beads, fluorescent and non-fluorescent. For given laser power, the channel number of the beads in a fluorescent histogram at a fixed amplifier setting should be recorded. If there is a decrease in the channel number of the fluorescent beads or an increase in the channel number for the non-fluorescent beads, the cleanliness of the flow cell and alignment of the instrument should be checked.

Laboratories undertaking routine clinical work should join a quality control scheme. Samples of cells are sent out regularly from a reference centre and each laboratory records the percentage of cells in certain sub-sets. The data from all the laboratories is collated and any laboratory whose data fall outside the normal range is notified. In the UK, a quality control scheme is run under the auspices of the Royal Microscopical Society from whom further details can be obtained (see Appendix C).

References

Camplejohn RS. (1994) The measurement of DNA content, alone or combined with other markers. In *Flow Cytometry: Clinical Applications* (ed. MG Macey). Blackwell Scientific, Oxford, pp. 215–236.

Clevenger CV, Shankey TV. (1993) Cytochemistry II. Immunofluorescence measurement of intracellular antigens. In *Clinical Flow Cytometry. Principles and Applications* (eds KD Bauer, RE Duque, TV Shankey). Williams and Wilkins, Baltimore, pp. 157–175.

Larsen JK. (1994) Measurement of cytoplasmic and nuclear antigens. In *Flow Cytometry. A Practical Approach*, 2nd edn. (ed. MG Ormerod). IRL Press, Oxford, pp. 93–117.

Polak JM, Van Noorden S. (1984) *An Introduction to Immunocytochemistry: Current Techniques and Problems*. Royal Microscopical Society Handbook no. 11. Oxford University Press, Oxford.

Schimenti KJ, Jacobberger JW. (1992) Fixation of mammalian cells for flow cytometric evaluation of DNA content and nuclear immunofluorescence. *Cytometry* **13**, 48–59.

Terstappen LWMM, Shah VO, Conrad MP, Rectenwald D, Loken MR. (1988) Discriminating between damaged and intact cells in fixed flow cytometric samples. *Cytometry* **9**, 477–484.

Further reading

Jackson AL, Warner NL. (1986) Preparation, staining and analysis by flow cytometry of peripheral blood leukocytes. In *Manual of Clinical Laboratory Immunology* (eds NR Rose, H Friedman, JL Fahey). American Society for Microbiology, Washington, DC, pp. 226–235.

Loken M, Wells D. (1994) Immunofluorescence of surface markers. In *Flow Cytometry. A Practical Approach*, 2nd edn. (ed. MG Ormerod). IRL Press, Oxford, pp. 67–91.

Lewis DE. (1993) Cytochemistry I: Cell surface immunofluorescence. In *Clinical Flow Cytometry. Principles and Applications* (eds KD Bauer, RE Duque, TV Shankey). Williams and Wilkins, Baltimore, pp. 143–156.

Radbruch A. (1992) Immunofluorescence: basic considerations. In *Flow Cytometry and Cell Sorting* (ed. A Radbruch). Springer-Verlag, Berlin, pp. 24–46.

Robinson JP. (ed.) (1993) *Handbook of Flow Cytometry Methods*. Wiley-Liss, New York.

Schwartz A, Fernandez-Repollet E. (1993) Development of clinical standards for flow cytometry. *Ann. NY Acad. Sci.* **677**, 28–39.

Stewart CC. (1990) Multiparameter analysis of leukocytes by flow cytometry. In *Flow Cytometry. Methods in Cell Biology*, Vol. 33 (eds Z Darzynkiewicz, HA Crissman). Academic Press, San Diego, pp. 427–450.

5 Analysis of DNA

5.1 Introduction

The second most common application of flow cytometry is the measurement of DNA to give a picture of the cell cycle and, in the case of clinical samples, to measure ploidy. Changes in the DNA histogram are used to study the mechanism of action of cytotoxic drugs since such compounds will generally disrupt the cell cycle. DNA analysis can also be combined with the measurement of antigen, an example being shown in *Figure 2.4*.

5.2 The cell cycle and the DNA histogram

A non-cycling cell is said to be in G0. For cycling cells, it is usual to define four distinct phases of the cycle (*Figure 5.1*). Cells commence the cycle in

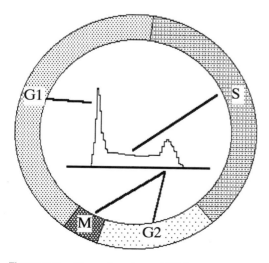

Figure 5.1:   The cell cycle and DNA content of the cell.

G1 in which normal cells have a diploid DNA content. Cells then enter the S (DNA synthetic) phase during which the DNA content increases until it has doubled, whereupon the cells are in G2. Eventually, cells enter mitosis (M phase) and divide to recommence their cycle.

By measuring the DNA content, it can be determined whether a cell is in G0/G1, S or G2/M phases of the cycle. Any change in cell cycle parameters (e.g. a slow down in S phase transit) will be reflected in the appearance of the DNA histogram.

The DNA histogram on its own gives a static picture of the cell cycle. Methods for a more dynamic measurement of cell cycle progression are discussed in Chapter 6.

5.3 Methods for staining DNA

There are a variety of fluorescent dyes which bind stoichiometrically to DNA (*Table 3.2*). These dyes are only weakly fluorescent in aqueous solution but fluoresce strongly when bound to DNA due to the hydrophobicity of their environment.

The most widely used dye is PI. It intercalates into double-stranded nucleic acids, is excited by the 488 nm line of an argon-ion laser and fluoresces red. Because it is excluded by viable cells, cells must be fixed or permeabilized or nuclei prepared before adding the dye. It also binds to double-stranded RNA, which should be removed first by treatment with RNase. Suitable protocols for use with PI are to be found in Ormerod (1994), Darzynkiewicz and Crissman (1990) and Radbruch (1992). The simplest method is to fix the cells in 70% ethanol. An alternative method is to enucleate the cells by suspending them in a buffer containing a detergent (Petersen, 1985).

Clinical specimens are often studied by extracting nuclei from routinely processed (that is, formalin-fixed, paraffin-embedded) tissue. Sections of 50 µm are cut from histological blocks and nuclei extracted, after removing the paraffin, by treatment with pepsin (Camplejohn, 1994; Hedley, 1990). The quality of the histograms obtained, which can be surprisingly good, depends on the way in which the tissue was handled initially in the histopathology laboratory.

To record a DNA histogram of viable cells, the bis-benzimidazole, Hoechst 33342, is used. Cells are incubated with the dye at 37°C. The rates of both uptake and egress of the dye vary from one type of cell to another. If the cells are under-stained, the DNA histogram will not be resolved. If the cells are over-stained, interaction between dye molecules causes fluorescence quenching and a red shift in the fluorescence emission spectrum (Watson, 1991). For each type of cell, preliminary experiments are needed to determine the narrow range of time and concentration of Hoechst 33342 over which a satisfactory DNA histogram will be obtained.

5.4 Quality control

During sample preparation, the aim is to obtain single particles, cells or nuclei, with minimum degradation of their DNA and minimum clumping. During the analysis, it is important that the profiles are as sharp as possible and that only single cells or nuclei are measured.

The quality of a DNA histogram is reflected in the width of the G0/G1 peak as measured by its cv (for a definition of 'cv', see Section 2.7.4). Small cvs give better resolution of small changes in ploidy in tumour samples and give a more reliable estimate of the different cell cycle components.

5.4.1 Standards

The instrument alignment is more critical for DNA measurement than for measurement of immunofluorescence. It should be checked frequently using a standard sample stained with the dye used for the experimental samples. Cells from a lymphoid organ (peripheral blood, bone marrow, thymus, tonsil) fixed and stored in 70% ethanol are suitable.

When studying cell cycle progression in cultured cells, no other standards are necessary. If the ploidy of tumour specimens is being measured, it is useful to have an absolute marker of DNA content. A sample can be split and human peripheral blood lymphocytes added to one half; this method has the disadvantage that double the size of sample is required. An alternative used by some workers is to dope the sample with nucleated erythrocytes (trout or chicken). The DNA content of these species is less than that of the human so that their histograms can be separated analytically from the human histogram. These types of marker are useful if fresh tissue is being handled and the standards are carried through the preparative procedures. They should not be used with paraffin-embedded tissues since the DNA–dye fluorescence will depend on the original fixation conditions used. In my experience, tumour samples always contain some stromal (normal) tissue whose DNA gives an inbuilt diploid marker. Any doubt about identification of the diploid peak can usually be resolved by inspecting the light scatter at 90° which is often higher in tumour nuclei (see *Figure 2.5*).

5.5 Excluding clumps of cells from the analysis

Analysis of the DNA histogram can be distorted by clumps of cells or nuclei, particularly two cells in G0/G1 of the cycle whose DNA content will equate to one cell in G2/M. Single cells can be resolved by analysing the shape of the fluorescent signal generated as a particle crosses the laser beam.

In instruments designed for DNA analysis, the laser beam is focused to give an elliptical cross-section whose width is close to, or less than, the diameter of a typical nucleus. As a particle crosses the beam, the integrated fluorescence will be proportional to the DNA content; the width of the signal in time will be the sum of the width of the particle and that of the laser beam. Because of the flow system, clumps of cells will tend to align along the direction of flow and will give a wider signal than single cells (*Figure 5.2*). The electronics in the flow cytometer have to be designed to analyse the shape of the fluorescent signal. Some instruments record the width of the signal vs. the integrated area, others the peak height vs. area. An example of the latter analysis is shown in *Figure 5.3*.

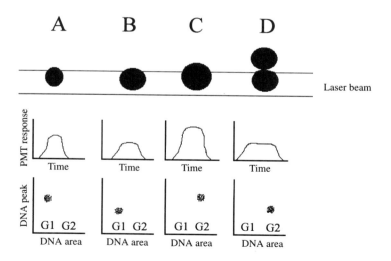

Figure 5.2: The shapes of the signals generated when different fluorescent particles cross a laser beam. The lower part of the figure shows an idealized cytogram of a peak vs. area plot of DNA fluorescence. A, a compact nucleus in G1; B, a normal nucleus in G1; C, a nucleus in G2; D, two G1 nuclei clumped together.

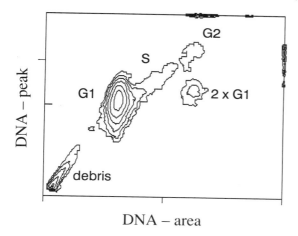

Figure 5.3:   A cytogram of peak vs. area DNA fluorescence showing clumped cells (2 × G1) with the same DNA content as cells in G2 but a lower peak. Fixed cells stained with PI and analysed on an Ortho Cytofluorograf, argon-ion laser at 488 nm.

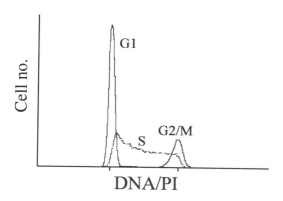

Figure 5.4:   The individual DNA histograms from cells in G1, S and G2 phases of the cell cycle. (Ortho Cytofluorograf, argon-ion laser at 488 nm.)

5.6 Computer analysis of the cell cycle phases

Figure 5.4 shows superimposed DNA histograms from G1, S and G2/M phases of the cycle for some rapidly cycling cells. The phases were separated analytically by pulse-labelling the S phase cells with BrdU (see Chapter 6). It can be seen that, because of instrumental broadening of the distributions, there is considerable overlap between early S phase and G1

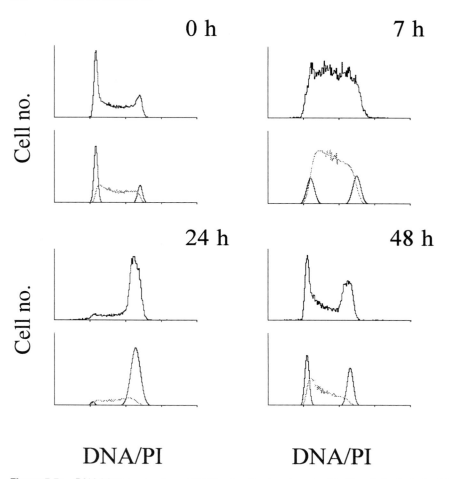

Figure 5.5:   DNA histograms from L1210 cells after treatment with 10 μM cisplatin for 2 h. The cells were fixed in 70% ethanol, washed and incubated with RNase and 10 μg ml⁻¹ PI in PBS. In each pair, the upper histogram shows the experimental data and the lower histogram shows a computer deconvolution into the phases of the cell cycle. The times after treatment are shown on the figure. (Ortho Cytofluorograf, argon-ion laser at 488 nm.)

and between late S phase and G2/M. The problem with analysis of a DNA histogram is finding a model to estimate reliably the extent of this overlap.

The variety of approaches for solving this problem have been summarized by Ormerod (1994) and Watson (1991). The most rigorous algorithms are probably the polynomial method of Dean and Jett (1974) and the 'pragmatic approach' of Watson (Ormerod *et al.*, 1987; Watson *et al.*, 1987). In practice, most researchers are restricted to the method packaged with the software on their instrument. Few algorithms will handle every histogram, particularly if the data is noisy or the cv large or if the cell cycle is severely distorted. The numbers generated should not be accepted blindly, but should be used in conjunction with the original DNA histogram. It should also be appreciated that the numbers produced by the computer programme are only estimates. Different algorithms will produce slightly different sets of numbers. However, changes in these numbers should consistently reflect changes in the cell cycle.

5.7 An example of analysis of the DNA histogram

Figure 5.5 shows some DNA histograms recorded from a murine lymphoma cell line, L1210, after treatment for 2 h with the cytotoxic drug, cisplatin. The cells were harvested at different times after treatment, fixed in 70% ethanol, centrifuged, resuspended in PBS with RNase and PI $(10 \ \mu g \ ml^{-1})$ and incubated at 37°C for 2 h. The Watson algorithm was used to deconvolute the histograms into the separate phases of the cell cycle.

After 7 h, there was a build-up of cells in S phase caused by a slow down in S phase transit. By 24 h, most of the cells had reached G2 where there

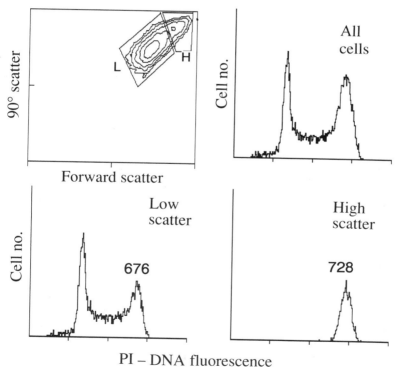

Figure 5.6: Light scatter and DNA recorded from L1210 cells 51 h after treatment with cisplatin. Two regions, of low (L) and high (H) light scatter, have been defined on the scatter plot. Their DNA histograms are shown together with the peak channel of G2.

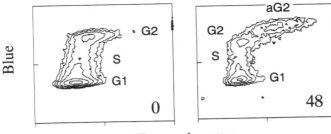

Figure 5.7:   Cytograms of DNA/Hoechst 33342 fluorescence vs. forward light scatter from unfixed L1210 cells 48 h after treatment with 10 μM cisplatin for 2 h. The cells were labelled by incubation with 10 μg ml^{-1} Hoechst 33342 at 37°C for 20 min. The numbers represent the time after treatment in hours. aG2 = abnormal G2 (increased forward light scatter and increased DNA content). (Ortho Cytofluorograf, argon-ion laser tuned to produce UV radiation.)

was a block. At 48 h, some of the cells had escaped the block and were cycling again. Quite a simple experiment yielded a wealth of information about the effect of cisplatin on these cells.

If light scatter is also recorded, more information can be derived. *Figure 5.6* shows that, at 48 h, there was an additional cluster in the cytogram of 90° vs. forward light scatter. This cluster consisted of cells blocked in G2 which had become abnormally large; their DNA content had also increased above that of normal cells in the G2/M phase of the cycle.

A similar conclusion can be drawn from a DNA histogram of unfixed cells (*Figure 5.7*). Cells were incubated in growth medium at 37°C for 20 min with Hoechst 33342 (10 μg ml^{-1}). After centrifugation, cells were resuspended in PBS containing 5 μg ml^{-1} PI. Dead cells which took up PI and fluoresced red were excluded from the analysis. Forty-eight hours after treatment, a cytogram of blue (Hoechst 33342) fluorescence vs. forward light scatter showed two populations – one with cells of normal size (as shown by forward light scatter) and a normal cell cycle; the other contained larger cells (increased forward light scatter) with a DNA content greater than that of normal cells in G2/M.

5.8 Measurement of the DNA histogram in combination with immunofluorescence

5.8.1 Surface antigens

Viable cells can be incubated with an antibody to a surface marker labelled with fluorescein, fixed in 70% ethanol and the DNA labelled with PI. Gating on green fluorescence will enable the DNA histograms of two sub-sets of cells to be displayed separately. This analysis can be extended

 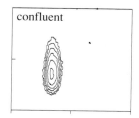 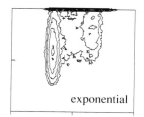

Red fluorescence – PI/DNA

Figure 5.8: A cell line derived from human breast, PMC42, fixed in 70% ethanol and stained with PI and for the nuclear protein, p53, using an FITC-labelled antibody. Cells at confluence and in the exponential phase of growth are shown. (Ortho Cytofluorograf, argon-ion laser at 488 nm.)

to two surface antigens by using a second label with PE. The cells must now be fixed with a low concentration of paraformaldehyde (0.25%) (PE does not survive fixation in ethanol) and permeabilized with a weak solution of detergent (0.2% Tween 20). PI has too great a spectral overlap with PE so the DNA is stained with 7-aminoactinomycin D (7-AAD) (Rabinovitch *et al.*, 1986; Schmid *et al.*, 1991). All three dyes are excited at 488 nm and the green (fluorescein), orange (PE) and deep red (7-AAD) fluorescences recorded together with 90° and forward light scatter.

5.8.2 Intracellular antigens

Intracellular antigens must be stained after fixation and permeabilization of the cells. The antigen of interest may be destroyed by some fixatives and the optimum fixation/permeabilization conditions must be determined for each particular antigen (see Section 4.3).

The antibody is labelled either directly or indirectly with FITC, the DNA labelled with PI and green versus red fluorescence recorded in the usual way (*Figure 5.8*).

References

Camplejohn RS. (1994) The measurement of DNA content alone or combined with other markers. In *Flow Cytometry: Clinical Applications* (ed. MG Macey). Blackwell Scientific, Oxford, pp. 215–236.

Darzynkiewicz Z, Crissman HA. (eds) (1990) *Flow Cytometry. Methods in Cell Biology,* Vol. 33. Academic Press, San Diego.

Dean PN, Jett JH. (1974) Mathematical analysis of DNA distributions derived from flow microfluorimetry. *J. Cell Biol.*, **60**, 523–527.

Hedley DW. (1990) DNA analysis from paraffin-embedded blocks. In *Flow Cytometry. Methods in Cell Biology,* Vol. 33 (eds Z Darzynkiewicz, HA Crissman). Academic Press, San Diego, pp. 139–147.

Ormerod MG. (1994) Analysis of DNA. General methods. In *Flow Cytometry. A Practical Approach*, 2nd edn. (ed. MG Ormerod). IRL Press, Oxford, pp. 69–87.

Ormerod MG, Payne AWR, Watson JV. (1987) Improved program for the analysis of DNA histograms. *Cytometry* **8**, 637–641.

Petersen SE. (1985) Flow cytometry of human colorectal tumours: nuclear isolation by detergent technique. *Cytometry* **6**, 452–460.

Rabinovitch PS, Torres RM, Engel D. (1986) Simultaneous cell cycle analysis and two-color surface immunofluorescence using 7-amino-actinomycin D and single laser excitation. *J. Immunol.* **136**, 2769–2776.

Radbruch A. (ed.) (1992) *Flow Cytometry and Cell Sorting*. Springer-Verlag, Berlin.

Schmid I, Uittenbogaart CH, Giorgi JV. (1991) A gentle fixation and permeabilisation method for combined cell surface and intracellular staining with improved precision in DNA quantification. *Cytometry* **12**, 279–285.

Watson JV. (1991) *Introduction to Flow Cytometry*. Cambridge University Press, Cambridge.

Watson JV, Chambers SH, Smith PJ. (1987) A pragmatic approach to the analysis of DNA histograms with a definable G1 peak. *Cytometry* **8**, 1–8.

6 Study of Cell Proliferation and Death

6.1 Introduction

A DNA histogram gives a static profile of the distribution of cells through the cell cycle. For example, a cell in S phase is not necessarily continuing to synthesize DNA. From *Figure 5.5*, it is not possible to determine whether the increase in the percentage of cells in G1 between 24 and 48 h is due to cells slowly overcoming the G2 block or whether they arise from a small number of normally cycling cells. More sophisticated methods are required to obtain a dynamic picture of the progression of cells through the cycle.

BrdU is incorporated into DNA by cells in place of thymidine. A detailed picture of cell proliferation can be obtained by measurement of the BrdU-labelled DNA. Two methods of detecting this label are described in the following sections.

BrdU can alter the behaviour of cells. Its presence can alter the nucleotide pools in a cell; it is a photo- and radiosensitizer and also enhances DNA damage caused by superoxide radicals. Checks should always be made using a simple DNA histogram to ensure that any effects observed are not modulated by the presence of BrdU.

The corollary to cell division is cell death. A complete picture of tissue growth and development can only be obtained by considering both proliferation and death. In Section 6.4, flow cytometric methods are described for measuring two forms of cell death, necrosis and apoptosis.

6.2 The BrdU/anti-BrdU method

6.2.1 Introduction

If cells are pulse-labelled with BrdU, those cells which are in S phase and have incorporated the precursor can be detected by an antibody to BrdU. Antibody labelling is correlated with the DNA histogram by the addition

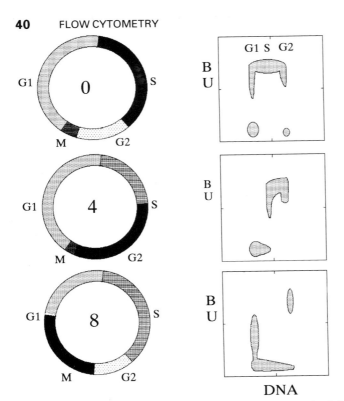

Figure 6.1:   The principle of the BrdU/anti-BrdU method for following cell cycle kinetics. A pulse label given at time 0 will move through the cell cycle as shown. This movement can be followed by recording cytograms of green (FITC–anti-BrdU) vs. red (PI–DNA) fluorescence. BU = anti-BrdU fluorescence.

of PI. By incubation of the cells after labelling, the subsequent movement of the labelled cells around the cycle can be followed (*Figure 6.1*).

Unfortunately, antibodies are not able to react with BrdU incorporated into DNA unless the structure of the chromatin is first disrupted. Disruption can be achieved by treatment with strong acid, heat or formamide or by controlled digestion with a nuclease. Suitable protocols have been given by Wilson (1994) (acid, formamide and heat), Beisker *et al.* (1987) (heat) and Dolbeare and Gray (1988) (nuclease). Care must be taken during the denaturation step since excessive treatment can adversely affect the DNA histogram.

The harsh treatments used to disrupt the DNA make it difficult to combine a study of cell kinetics with measurement of an antigen, particularly a nuclear protein. If this is desired, the most reliable procedure is to use a nuclease in the denaturation step (Toba *et al.*, 1992). If FITC and PE are used to label the two antibodies, then 7-AAD should be used to stain the DNA (see Section 5.8.1).

Some monoclonal antibodies show specificity for one of the halogeno-pyrimidines so that additional information can be gained by pulse-labelling at a later time with a second precursor, either chloro- or iododeoxyuridine (Bakker *et al.*, 1991; Dolbeare *et al.*, 1988). This approach allows two time points to be obtained from a single sample which is particularly useful for clinical studies on human tumours *in vivo*.

6.2.2 Data obtained from cultured cells

Figure 6.2 shows data from a normally cycling cell line (Chinese hamster V79). Immediately after labelling, the anti-BrdU had labelled the cells in S phase; G1 and G2/M phases were unlabelled. Three hours later, a few of the labelled cells had moved through G2, divided and were in G1. The rest of the labelled cells were in G2 and mid to late S phase. There were few unlabelled cells in G2 and the unlabelled cells originally in G1 were now progressing through to S phase. With time, the number of labelled cells in G1 increased and unlabelled cells could be observed moving through S phase. After 8 h, labelled cells were again moving into S phase. A simple

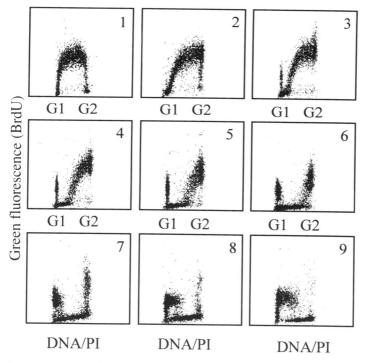

Figure 6.2:   Chinese hamster, V79, cells were incubated in suspension culture for 30 min with 10 µM BrdU. The cells were washed and incubated further at 37°C for the times indicated, after which the cells were harvested and fixed in 70% ethanol. They were washed in PBS and their DNA disrupted by incubation in 2 M HCl at room temperature for 30 min. After washing to remove acid, they were incubated for 1 h with a rat monoclonal anti-BrdU antibody (ICR2, SeraLab, Crawley Down, UK), washed and finally incubated with FITC-labelled rabbit anti-rat Ig (Sigma) for 30 min, washed and 10 µg ml^{-1} PI added. In the flow cytometer, cells were excited at 488 nm and green (FITC) and red (DNA) fluorescences recorded. Data recorded on a Becton-Dickinson FACScan; excitation at 488 nm. Redrawn from Wilson (1994) with permission from Oxford University Press.

visual inspection of the data shows that the cell cycle time for these cells was about 11 h, at which time the original pattern would have been re-established.

There are several ways of analysing these data. One method is to measure the rate of movement of cells through S phase ('relative movement') from which the time spent by a cell in S phase (T_S) can be calculated. This measurement can be made from a single time point which enables T_S to be calculated from a tumour labelled *in vivo* (Section 6.2.3 below) (Begg *et al.*, 1985). A detailed discussion of this analysis has been given by Terry *et al.* (1991).

The cell cycle time is calculated from the formula: $\lambda \times T_S/LI$, where LI is the labelling index (percent labelled cells at time 0) and λ is a factor (usually taken to be 0.8) to correct for the age distribution through the cell cycle. The rate of appearance of labelled cells in G1 also enables T_{G2+M} to be calculated. When calculating LI and T_{G2+M}, a correction must be applied to allow for the increase in the number of labelled cells when they divide.

An alternative method of analysis is to place a window in part of the cell cycle (e.g. in the centre of S phase) and to measure the rate of transit of labelled cells through the window. This method is equivalent to that of percent labelled mitoses (plm) in which cells are pulse labelled with [^{3}H]thymidine, fixed for autoradiography at later times and the percentage of labelled mitoses vs. time is recorded.

6.2.3 Data obtained from tumours labelled *in vivo*

The half-life of BrdU in an animal is about 20 min, so that a single injection of the nucleoside is equivalent to a pulse label. The tumour is usually biopsied 4–6 h later and the tissue chopped and fixed in ethanol for processing at a later time. *Figure 6.3* shows data obtained from a rat mammary tumour. There was a 4 h gap between injection of the BrdU and removal of the tumour. The 'window' of unlabelled cells in early S phase is equivalent to 4 h of S phase. From the length of this 'window', a crude calculation of T_S can be made; this estimate can be a useful check on the value derived from a more sophisticated calculation based on the 'relative movement' (Begg *et al.*, 1985).

6.3 BrdU–Hoechst/PI method for measuring cell cycle kinetics *in vitro*

6.3.1 Introduction

The fluorescence of bis-benzimidazole dyes (Hoechst 33258 and Hoechst 33342) bound to DNA is quenched by BrdU. Consequently, continuous

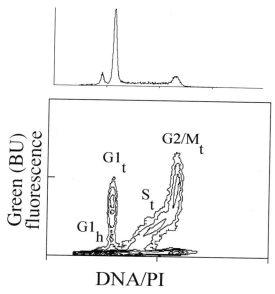

Figure 6.3: Rat mammary tumour cells labelled *in vivo* with BrdU. Nuclei were then labelled with anti-BrdU and PI as described in *Figure 6.2*. The subscript 'h' designates the normal, diploid, host and 't' the aneuploid tumour cells. A DNA histogram is also shown in which the peak from the diploid cells can be distinguished. Cells prepared by Dr T.A.D. Smith and data recorded by Mrs J. Titley on an Ortho Cytofluorograf with an argon-ion laser tuned to 488 nm.

labelling with BrdU and subsequent staining of DNA with Hoechst 33258 separates cells according to the number of replications they underwent during the period of labelling. Addition of a DNA label unaffected by BrdU, such as PI, resolves the cell cycle into the G0/G1, S and G2/M compartments (Rabinovitch *et al.*, 1988). In this method, cultured cells are incubated with BrdU for different times, the cells harvested, nuclei prepared by lysis with a detergent (0.1% Nonidet-P40) and stained by the addition of Hoechst 33258 and either PI or ethidium bromide (EB). In the flow cytometer, the cells are excited with UV light and red (PI–DNA) vs. blue (Hoechst–DNA) fluorescence is recorded (*Figure 6.4*).

Many studies have reported the analysis of the rate of exit from the G0/G1 compartment of the cell cycle using cells which have been synchronized at this stage (Poot *et al.*, 1990 and references therein). The method is not limited to synchronous cells and it has also been used to analyse cell cycle progression in asynchronous cultures (Ormerod and Kubbies, 1992, and references therein). Both applications are described in this chapter.

6.3.2 Some general considerations (see also Poot and Ormerod, 1994)

There are several precautions which should be taken when cells are labelled continuously with BrdU. Different cells have different thymidine pool sizes, which affect the rate of incorporation of BrdU. For each type of cell, the lowest concentration of BrdU which gives sufficient quenching of the Hoechst fluorescence must be determined in a preliminary experiment. Concentrations between 10 µM (human ovarian carcinoma cell line) and 100 µM (human peripheral blood lymphocytes) BrdU have been used. The growth of the cells in the presence of BrdU and their cell cycle (by conventional staining, see Chapter 4) should be checked. If growth is reduced and/or cells accumulate in G2, it may be necessary to lower the BrdU concentration and/or add an equimolar concentration of deoxy-cytidine.

The plating density of the culture should be sufficiently low so that the BrdU is not exhausted and so that the cells do not reach plateau phase of growth during the experiment. Typical starting concentrations are 3×10^3 cells/cm^2 for adherent cells or 2×10^5 cells/ml for suspension cultures. BrdU is a photo- and radiosensitizer and, during and after labelling, cell cultures should be protected from light.

In the arc lamp-based instruments and in instruments using an argon-ion laser as a source of UV, PI is excited via resonance energy transfer from the Hoechst 33258 (for an explanation of energy transfer, see Section 3.2). This effect results in some quenching of the red fluorescence in the presence of BrdU. He–Cd lasers have an output at 325 nm which excites EB and PI directly, giving better resolution of the cell cycle (Kubbies *et al.*, 1992).

6.3.3 Synchronized cells

These experiments generally use either naturally quiescent lymphocytes from peripheral blood, spleen or thymus or cultured cells rendered quiescent by reduction of growth factors (frequently fibroblasts cultured in the absence of serum). The BrdU is added at the same time, as a factor to stimulate the cells to cycle – for example, phytohaemagglutinin for lymphocytes or serum for fibroblasts. If a cytotoxic agent is being studied, the drug is usually also added at the start of the experiment.

Figure 6.4 shows a bivariate cytogram obtained from human peripheral blood lymphocytes 72 h after stimulation with phytohaemagglutinin. There is a cluster of cells which remained quiescent and had not left G0/G1. The signal track moving from the G0/G1 cluster to the left represented cells in the first S phase culminating in G2. The G1' cluster are cells which traversed a full cell cycle and underwent mitosis. Cell division caused halving of the fluorescence intensity of the cell in the first G2 in both the Hoechst and PI direction. The track from cells in the second S phase (S') moved to the right because there was less quenching on the second round of labelling. A cluster of cells which had divided twice (G1'') can be resolved.

6.3.4 Asynchronous cultures

BrdU is normally added to cultures (suspension or adherent) of cells which have been established in the early stages of exponential growth. If the

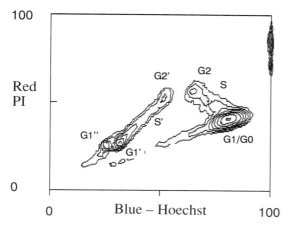

Figure 6.4: The BrdU–Hoechst/PI method applied to synchronized cells. A bivariate cytogram of red (PI) vs. blue (Hoechst) fluorescence from human peripheral blood lymphocytes 72 h after stimulation with phytohaemagglutinin and simultaneous addition of 80 μM BrdU. After 72 h, the cells were suspended in 100 mM Tris–HCl (pH 7.4), 154 mM NaCl, 1 mM CaCl$_2$, 0.5 mM MgCl$_2$, 0.1% Nonidet–P40, 1.2 μg ml^{-1} Hoechst 33258 and 2 μg ml^{-1} PI. Data were recorded on an Ortho Cytofluorograf with an argon-ion laser tuned to give UV. G1, S, G2, the first; G1′, S′, G2′, the second, and G1″, the third cell cycle. Further details are given in the text.

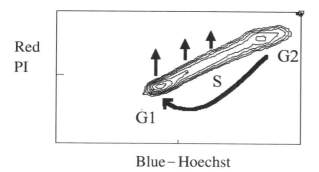

Figure 6.5: The BrdU–Hoechst/PI method. A bivariate cytogram of red (PI) vs. blue (Hoechst) fluorescence from cells before the addition of BrdU. The arrows indicate the initial movement of cells in the cytogram after the addition of BrdU.

effects of radiation (high energy, UV or heat) are to be studied, the cultures are treated before adding BrdU. Short-term drug treatments are also carried out before adding BrdU; for continuous treatment with a cytotoxic drug, the compound is normally added with the BrdU. Extra information about the proliferative fate of the cells can sometimes be obtained by adding the BrdU some time after the drug (Ormerod *et al.*, 1994).

Figure 6.5 shows diagrammatically how the different populations of cells 'track' in the cytogram of red vs. blue fluorescence. *Figure 6.6* shows experimental data obtained from an exponentially growing suspension culture of a murine lymphoma cell line (L1210). Initially, both red and blue

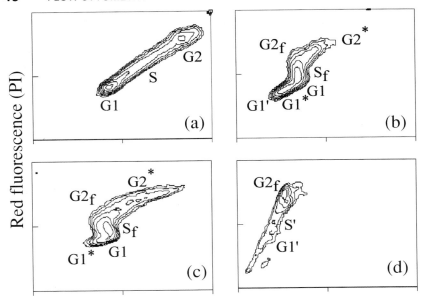

Blue fluorescence (Hoechst)

Figure 6.6: The BrdU–Hoechst/PI method. Bivariate cytograms of red (PI) vs. blue (Hoechst) fluorescence from a murine lymphoma cell line (L1210); 50 μM BrdU was added at time 0. (a) and (b): cells under normal growth conditions after 0 and 7 h respectively; (c) and (d): cells after 10 μM cisplatin for 2 h after further incubation for 8 and 24 h, respectively. For conditions of staining the cells and recording the data, see the legend to *Figure 6.4*. G1, S, G2 – initial cell cycle; S_f, $G2_f$, $G2^*$ – first cycle cells originating from G1 (S_f, $G2_f$) and S ($G2^*$) phases; G1′, $G1^*$, S′ – second cycle cells originating from G1 (G1′, S′) and S ($G1^*$) phases. Data recorded on an Ortho Cytofluorograf with an argon-ion laser tuned to give UV.

fluorescence gave a normal cell cycle with cells in G1, S and G2/M (panel (a)). In panel (b), cells after 7 h incubation are shown. Cells had disappeared from G2 due to cell division. Cells in late to mid S at time 0 had divided and are to be found in G^*. Cells in early S phase at time 0 have taken up BrdU so that, while their red fluorescence increased in line with DNA content, their quenched blue fluorescence did not increase and they were now in $G2^*$. Cells which had moved from G1 into S phase are labelled S_f and those which had reached G2, $G2_f$. Cells that were in G1 at time 0, after they had completed traverse of the cell cycle, would be found at G1′.

Panels (c) and (d) show cells 8 and 24 h after treatment with 10 μM cisplatin for 2 h. Cells in G2 at the time of treatment had divided. There had been a slow down in S phase transit (see also *Figure 5.5*). Cells from late S had reached G2 ($G2^*$) but cells from early and mid S were still in S phase. There was also an accumulation of cells in S from G1 (S_f). Few cells in S phase at the time of treatment had divided and reached $G1^*$. At 24 h, the cells which were in G1 at time 0 were to be found blocked in G2 ($G2_f$). The cells which were in S and G2 at time 0 had now divided and were cycling (G1' and S'). A G2 block can be observed in a simple DNA histogram (*Figure 5.5*) but the BrdU-Hoechst/PI method reveals additional information. The cells in $G2_f$ were in G1 at the start of the experiment; cells which were in S phase and G2 phase at the time of addition of the drug have divided. The deduction can be made that cells are only blocked in G2 after replication of DNA on a damaged template (Ormerod *et al.*, 1994).

Because of the complexity of the pattern observed with asynchronous cells, it is usually not possible to observe more than one cell cycle, although sometimes cells in a second G1 (G1'') can be seen.

Numerical data can be obtained by drawing regions around the different cell cycle compartments (Ormerod and Kubbies, 1992). Because of the overlap between the G1 and early S phases and G2 and late phases, this method will overestimate the G phases and underestimate S phases. Unfortunately, it is difficult to avoid this source of error without recourse to complex computer programs. The data should be corrected to take into account that, as cycling cells divide, they dilute any undivided cells in the culture.

6.3.5 Combined antibody staining and BrdU–Hoechst/PI analysis

If the method is to be combined with staining for a surface antigen, the cells are incubated with a fluorescein-labelled antibody and then fixed before suspending in the staining buffer. Two lasers have to be used – one tuned to 488 nm to excite the fluorescein and (optionally) the PI, the other tuned to output UV to excite the Hoechst–DNA complex.

If two antibodies are used, one should be labelled with fluorescein and the other with PE. Because of the spectral overlap between PE and PI, 7-AAD is used as a DNA label.

6.4 Quantification of cell death

6.4.1 Estimating cell viability

The number of dead cells in a suspension is often estimated by counting the cells which take up an acidic dye such as trypan blue. A similar method can be used on the flow cytometer by adding PI to the cells. PI is excluded by viable cells and, when taken up by dead or dying cells, binds to nucleic acids and fluoresces red (see Chapters 3 and 5). The use of flow cytometry has the advantage that large numbers of cells can be counted quickly and that the determination of negative/positive is objective.

The resolution of the method can be improved by the additional use of fluorescein diacetate (FDA) or one of the related compounds. FDA, which is not fluorescent, is taken up by cells in which it is converted to fluorescein

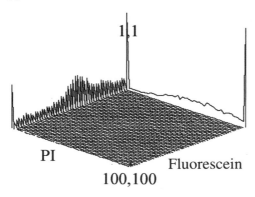

Figure 6.7:   Cells from a human bladder cell line (U1) stained with FDA and PI. A cytogram of green versus red fluorescence showing viable (green positive, red negative) and dead (green negative, red positive) cells. Cells prepared by Ms J. Mills and recorded by Mrs J. Titley, Institute of Cancer Research, on an Ortho Cytofluorograf. Excitation at 488 nm.

by the intracellular esterases (see Section 3.4.3). Fluorescein is retained by the cell if the plasma membrane is intact. A typical result is shown in *Figure 6.7*. In the cytogram of red vs. green fluorescence, the viable cells are counted as those which are positive for green and negative for red fluorescence. There may also be an intermediate population which shows both weak green and red fluorescence. These are probably dying cells and should not be counted in the viable population.

Dead cells usually show decreased forward light scatter and slightly increased orthogonal scatter. Sometimes light scatter on its own is sufficient to distinguish dead and live cells. If either PI or PI plus FDA is used, then the light scatter gate should be set generously to include all the dead and live cells but to exclude clumps of cells and debris.

Flow cytometric methods measure cells with an intact plasma membrane and 'dead' cell is taken to mean a cell with a damaged plasma membrane. The terminology can be slightly misleading since cells scored as 'viable' may not be capable of division – for example after a lethal dose of high energy radiation.

6.4.2 Measurement of apoptotic cells

Introduction. Two distinct modes of cell death – apoptosis and necrosis – have been recognized in eukaryotic cells. Apoptosis can be regarded as a form of cell suicide in that, once the process has been triggered, a cell is destroyed by a sequence of events driven from within the cell. It is characterized morphologically by condensation of nuclear chromatin, compaction of cytoplasmic organelles, cell shrinkage and changes at the cell surface. At a late stage in apoptosis, the dying cell is phagocytosed by neighbouring cells. During apoptosis *in vitro*, rupture of the plasma membrane occurs only at a late stage, as opposed to necrosis during which damage to the plasma membrane is an earlier event. Often apoptosis is accompanied by fragmentation of DNA into oligonucleosomal fragments with lengths which are multiples of 180–200 bp. DNA degradation in necrotic cells tends to be non-specific. More detailed descriptions of apoptosis can be found in Arends and Wyllie (1991) and Wyllie (1993).

Three basic methods for detecting apoptotic cells by flow cytometry have been published (Darzynkiewicz *et al.*, 1992; Gorczyca *et al.*, 1992; Ormerod *et al.*, 1992, 1993a). The first measures changes in the DNA content due to DNA degradation, the second labels the ends in DNA caused by endonucleolytic action and the third measures a change in the permeability of the plasma membrane.

Detecting apoptotic cells by measuring a DNA histogram. Frequently, at a late stage in apoptosis, double-strand breaks are introduced into the DNA at the linker regions between the nucleosomes. During fixation of the cells, some of the lower molecular weight fragments leach out, lowering the DNA content. These cells can be observed on a DNA histogram as a hypoploid or 'sub-G1' peak (Darzynkiewicz *et al.*, 1992, and references therein).

Typical DNA histograms are shown in *Figure 6.8*.

Labelling ends of cut DNA. Strand breaks in the DNA can be measured in fixed cells by adding labelled nucleotides enzymatically on to the cut ends either by nick translation using a DNA polymerase or by using a terminal deoxynucleotidyl transferase (TdT) (Gorczyca *et al.*, 1993). Labels used include biotin, in which case the biotinylated DNA is then visualized by reaction with fluorescein-labelled avidin, and digoxigenin followed by fluorescein-labelled anti-digoxigenin. The bulk DNA is stained

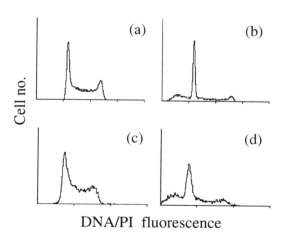

Figure 6.8: DNA histograms of cultures of a murine lymphoma cell line (L1210) undergoing apoptosis. The cultures were incubated for 16 h in either full growth medium (control) ((a) and (c)) or arginine-deficient medium (which induced apoptosis) ((b) and (d)). (a, b): cells fixed in 70% ethanol and stained with PI. (c, d): cells suspended in 100 mM Tris–HCl (pH 7.4), 154 mM NaCl, 1 mM CaCl$_2$, 0.5 mM MgCl$_2$, 0.1% Nonidet–P40, 4 µg ml^{-1} PI. Data recorded on an Ortho Cytofluorograf with an argon-ion laser tuned to 488 nm.

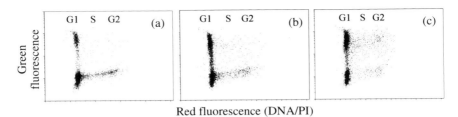

Figure 6.9:   *In situ* end labelling of DNA strands in apoptotic cells. Immature rat thymocytes incubated with (a) no drug, (b) 0.1 µM dexamethasone or (c) 10 µM etoposide for 4 h at 37°C. Cells were suspended in 100 µl TdT reaction buffer (100 mM Na cacodylate, 10 mM cobalt chloride, 1 mM dithiothreitol, pH 7.2, containing 10 units TdT, 2 µM digoxigenin-11-dUTP) and incubated at 37°C for 30 min. After washing, cells were labelled with anti-digoxigenin–fluorescein FAB fragments (30 min at room temperature), washed and resuspended in PBS with 10 µg ml^{-1} PI. Cytograms were recorded on a Becton-Dickinson FACScan employing a 15 mW argon-ion laser tuned to 488 nm and analysed using PCLYSIS software. Data supplied by James Wolfe and Dr Gerry M. Cohen, MRC Toxicology Unit, University of Leicester.

with PI. Apoptotic cells can be enumerated in a cytogram of green (fluorescein) vs. red (PI) fluorescence. This method has the advantage that apoptotic cells throughout the cell cycle can be identified. An example of its application is shown in *Figure 6.9*. In culture, thymocytes undergo apoptosis spontaneously (panel a); apoptosis is accelerated by dexamethasone (panel b) and etoposide (panel c). The figure shows that, while spontaneous and dexamethasone-induced apoptosis occurred mainly from cells in G0/G1 of the cell cycle, etoposide additionally induced apoptosis in cells in S and G2/M phases (J. Wolfe and G.M. Cohen, personal communication).

The Hoechst–PI method.   The permeability of the plasma membrane of cells alters during apoptosis. As a result, cells may take up the dye, Hoechst 33342, more rapidly (Ormerod *et al.*, 1992, 1993b). 'Dead' cells may be distinguished by adding PI which will be excluded by cells with an intact plasma membrane (see above). This method uses unfixed cells so that it is possible to enumerate normal, apoptotic and 'dead' cells in the same assay. Viable cells can also be sorted for further study.

Figure 6.10 shows a result obtained from immature rat thymocytes induced to undergo apoptosis *in vitro* by treatment with dexamethasone. The cluster from the apoptotic cells, whose identity has been verified by electron microscopy of sorted cells (Sun *et al.*, 1992), is quite distinct from that from normal cells. The apoptotic thymocytes show reduced forward angle light scatter while other types of cell may show increased forward light scatter (Ormerod *et al.*, 1993a). Reduction in size of the apoptotic cells reduces the light scatter over a narrow angle in the forward direction; changes in chromatin structure increase light scatter, particularly at wider angles. The difference in light scatter between normal and apoptotic cells is determined by the cell type and the angle over which forward light scatter is collected.

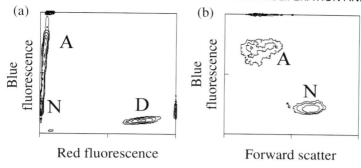

(a) Blue fluorescence vs Red fluorescence; A, N, D

(b) Blue fluorescence vs Forward scatter; A, N

Figure 6.10: Immature rat thymocytes treated with 0.1 μM etoposide for 4 h and then incubated with 1 μg ml^{-1} Hoechst 33342 for 8 min at 37°C followed by the addition of 5 μg ml^{-1} PI. (a): cytogram of blue (Hoechst–DNA) vs. red (PI–DNA) fluorescence. (b): cytogram of blue fluorescence vs. forward light scatter of the viable cells (those stained with PI have been excluded in the computer analysis). N, normal; A, apoptotic; D, dead cells. Note that the apoptotic cells have reduced light scatter. Cells were prepared by Xiao-Ming Sun and data recorded by Roger Snowden, MRC Toxicology Unit, Leicester, on an Ortho Cytofluorograf with a krypton laser tuned to give UV.

The rate of uptake of Hoechst 33342 into cells depends on the cell type, and the optimum time of incubation with the dye to give the best separation between normal and apoptotic cells must be determined by experiment. Since the method measures a change in the plasma membrane of the apoptotic cells (Ormerod *et al.*, 1993b), it should always be validated before routine use. Some drugs which induce apoptosis might also affect the plasma membrane.

References

Arends MJ, Wyllie AH. (1991) Apoptosis: mechanisms and role in pathology. *Int. Rev. Exp. Pathol.* **32**, 223–254.

Bakker PJM, Stap J, Tukker CJ, van Oven CH, Veenhof CHN, Aten J. (1991) An indirect immunofluorescence double staining procedure for the simultaneous flow cytometric measurement of iodo- and chlorodeoxyuridine incorporated into DNA. *Cytometry* **12**, 366–372.

Begg AC, McNally NJ, Shrieve DC, Karcher H. (1985) A method to measure the duration of DNA synthesis and the potential doubling time from a single sample. *Cytometry* **6**, 620–626.

Beisker W, Dolbeare F, Gray JW. (1987) An improved immunochemical procedure for high sensitivity detection of incorporated bromodeoxyuridine. *Cytometry* **8**, 235–239.

Darzynkiewicz Z, Bruno S, Del Bino G, Gorczyca W, Hotz MA, Lassota P, Traganos F. (1992) Features of apoptotic cells measured by flow cytometry. *Cytometry* **13**, 795–808.

Dolbeare F, Gray JW. (1988) Use of restriction endonucleases and exonuclease III to expose halogenated pyrimidines for immunochemical staining. *Cytometry* **9**, 631–635.

Dolbeare F, Kuo WL, Vanderlaan M, Gray JW. (1988). Cell cycle analysis by flow cytometric analysis of the incorporation of iododeoxyuridine (IdUrd) and bromodeoxyuridine (BrdUrd). *Proc. Am. Assoc. Cancer Res.* **29**, 1896–1901.

Gorczyca W, Bruno S, Darzynkiewicz RJ, Gong J, Darzynkiewicz Z. (1992). DNA strand breaks occurring during apoptosis: their early *in situ* detection by the terminal deoxynucleotidyl transferase and nick translation assays and prevention by serine protease inhibitors. *Int. J. Oncol.* **1**, 639–648.

Kubbies M, Goller B, Van Bockstaele DR. (1992). Improved BrdU–Hoechst bivariate cell kinetic analysis by helium–cadmium single laser excitation. *Cytometry* **13**, 782–786.

Ormerod MG, Kubbies M. (1992) Cell cycle analysis of asynchronous cell populations by flow cytometry using bromodeoxyuridine label and Hoechst–propidium iodide stain. *Cytometry* **13**, 678–685.

Ormerod MG, Collins MKL, Rodriguez-Tarduchy G, Robertson D. (1992) Apoptosis in interleukin-3-dependent haemopoetic cells. Quantification by two flow cytometric methods. *J. Immunol. Methods* **153**, 57–65.

Ormerod MG, Sun X-M, Brown D, Snowden RT, Cohen GM. (1993a) Quantification of apoptosis and necrosis by flow cytometry. *Acta Oncol.* **22**, 417–424.

Ormerod MG, Sun X-M, Snowden RT, Davies R, Fearnhead H, Cohen GM. (1993b) Increased membrane permeability of apoptotic thymocytes: a flow cytometric study. *Cytometry* **14**, 595–602.

Ormerod MG, Orr RM, Peacock JH. (1994) The role of apoptosis in cell killing by cisplatin: a flow cytometric study. *Br. J. Cancer* **69**, 93–100.

Poot M, Ormerod MG. (1994) Analysis of cell proliferation using continuous bromodeoxyuridine labelling and bivariate Hoechst 33258/ethidium bromide flow cytometry. In *Flow Cytometry. A Practical Approach*, 2nd edn. (ed. MG Ormerod). IRL Press, Oxford, pp. 157–167.

Poot M, Kubbies M, Hoehn H, Grossman A, Chen Y, Rabinovitch PS. (1990) Cell cycle analysis using continuous bromodeoxyuridine labelling and Hoechst 33258–ethidium bromide bivariate flow cytometry. In *Methods in Cell Biology*, Vol. 33 (eds Z Darzynkiewicz, HA Crissman). Academic Press, New York, p. 186.

Rabinovitch PS, Kubbies M, Chen YC, Schindler D, Hoehn H. (1988) BrdU–Hoechst flow cytometry. A unique tool for quantitative cell cycle analysis. *Exp. Cell Res.* **174**, 309–318.

Sun X-M, Snowden RT, Skilleter DN, Dinsdale D, Ormerod MG, Cohen GM. (1992) A flow cytometric method for the separation and quantitation of normal and apoptotic thymocytes. *Anal. Biochem.* **204**, 351–356.

Terry NHA, White RA, Meistrich ML, Calkins DP. (1991) Evaluation of flow cytometric methods for determining population potential doubling times using cultured cells. *Cytometry* **12**, 234–241.

Toba K, Winton EF, Bray RA. (1992) Improved staining method for the simultaneous flow cytofluorometric analysis of DNA content, S-phase fraction, and surface phenotype using single laser instrumentation. *Cytometry* **13**, 60–67.

Wilson GD. (1994) Analysis of DNA-measurement of cell kinetics by the bromodeoxyuridine/anti-bromodeoxyuridine method. In *Flow Cytometry. A Practical Approach*, 2nd edn. (ed. MG Ormerod). IRL Press, Oxford, pp. 137–156.

Wyllie AH. (1993) Apoptosis. *Br. J. Cancer* **67**, 205–208.

Further reading

Gray JW, Mayall BH. (eds) (1985) *Monoclonal Antibodies Against Bromodeoxyuridine.* Alan R. Liss, New York.

Gray JW, Darzynkiewicz Z. (eds) (1987) *Techniques in Cell Cycle Analysis.* Humana Press, Clifton, NJ.

Ormerod MG. (ed.) (1994) *Flow Cytometry. A Practical Approach*, 2nd edn. IRL Press, Oxford.

7 Other Applications

7.1 RNA content

DNA and RNA content can be measured in permeabilized cells simultaneously by labelling with acridine orange (AO) (Darzynkiewicz and Kapuscinski, 1990; see Section 3.4.2). Excitation is at 488 nm; the DNA–AO complexes fluoresce green and the RNA–AO red. This method has been used to distinguish sub-compartments within the G1 phase of the cell cycle. Unfortunately it is not simple to apply. The concentration of AO and cells used is critical; the results can also be affected by the geometry of the flow system of the cytometer.

Pyronin Y has been used to stain RNA after the binding to DNA has been blocked either with Hoechst 33342 or methyl green (Pollack *et al.*, 1982). If a two-laser instrument is available, a combination of pyronin Y and Hoechst 33342 can be used for the combined measurement of RNA and DNA (Crissman *et al.*, 1985).

Reticulocytes in whole blood can be enumerated by flow cytometry by staining the RNA of unfixed cells with thiazole orange (Lee *et al.*, 1986; see Section 3.3.2). Binding of thiazole orange to DNA is unimportant since reticulocytes do not have a nucleus. The number of reticulocytes in blood is an important diagnostic indicator of erythropoetic activity.

7.2 Protein content

Individual proteins are usually detected using antibodies. The total protein content of a cell may be estimated by staining fixed cells, either with a fluorescent protein stain, such as sulphorhodamine 101, or with a probe which binds covalently to proteins, such as FITC (Crissman *et al.*, 1985). These measurements are not commonly made, possibly because a change in protein content is often related to change in cell size which can be detected by light scatter.

The protein content of isolated nuclei has been measured using reaction with FITC (Pollack, 1990). Under certain conditions of staining, the availability of lysine residues for reaction with the FITC may be affected by the structure of the chromatin so that the amount of stain can reflect the chromatin structure (Dyson *et al.*, 1989).

7.3 Kinetic analysis of intracellular enzymes

The conversion of an ester of fluorescein (such as FDA, see Section 3.4.3, *Figure 3.2*) to its fluorescent form can, in itself, be used as a measure of intracellular esterase activity. Other derivatives of fluorescein, of 4-methylumbilliferone and of α-naphthol have been used to measure a variety of enzymes. Valet and co-workers have produced novel peptide substrates linked to rhodamine for the analysis of proteases (Assfalg-Machleid *et al.*, 1992; Rothe *et al.*, 1992). Watson (1991) has provided an extensive discussion of the analysis of enzyme kinetics by flow cytometry.

An interesting application of these methods is the detection of the presence of the *Escherichia coli* β-D-galactosidase gene (*lacZ*) in mammalian cells (Fiering *et al.*, 1991; Nolan *et al.*, 1988). The *lacZ* gene is used as a 'reporter' gene during the introduction of cloned DNA constructs into cultured cells. Galactosidase is detected by the production of fluorescein from the substrate, fluorescein β-galactopyranoside. Individual cells which have incorporated the foreign construct can then be sorted.

7.4 Membrane permeability

When a cell is loaded with a fluorescent compound, such as fluorescein, the reporter molecules, although charged, eventually leak out of the cell. Measurement of the loss of fluorescence with time after loading can be used to monitor changes in membrane permeability. Derivatives of fluorescein with different carboxy and chlorine substituents have different rates of loss of fluorescence from cells. The compound which shows the optimum difference between control and test samples should be selected.

7.5 Membrane potential

There are a several lipophilic dyes which partition between the cell and the surrounding medium according to the plasma membrane potential. Of particular use are a series of cyanine dyes which, depending on their substituents, have different wavelengths of excitation and emission (Waggoner, 1985). These dyes are positively charged and may be taken up by mitochondria. This complication can be avoided by using oxonol dyes (Chused *et al.*, 1986).

Rhodamine 123 is a membrane potential probe which stains mitochondria brightly (Weiss and Chen, 1984). Its partitioning between the medium and cytoplasm and the cytoplasm and mitochondria is complex and has yet to be explored fully.

7.6 Production of intracellular oxidative species

The reduced form of 2′, 7′-dichlorofluorescein (dichlorofluorescin, DCFH), which is not fluorescent, can be loaded into cells as an ester. After conversion of the ester to DCFH by intracellular esterases, oxidation will yield a fluorescent product. DCFH has been used to observe oxidative burst in neutrophils (Bass *et al.*, 1983) and also the production of intracellular oxidative species by ionizing radiation (Ormerod and Peacock, unpublished observations).

7.7 Measurement of drug uptake

DNA is the site of action of many chemotherapeutic drugs. The uptake and retention of drugs which are fluorescent, in particular anthracyclines such as adriamycin and daunomycin, can be measured by flow cytometry (Krishan *et al.*, 1987). Both these dyes can be excited at 488 nm and they fluoresce orange.

Resistance to a whole series of drugs (multi-drug resistance) is related to the rapid efflux of drugs from the cell mediated by the *p*-glycoprotein pump. The DNA-binding dye, Hoechst 33342, is probably pumped out of the cell by the same mechanism and this dye has been used to study the effect of drugs which inhibit efflux (e.g. the calcium channel blocker, verapamil) (Krishan, 1987). The dye can also be used to define functionally the multi-drug resistance in tumour cells (Watson, 1991).

7.8 Binding and endocytosis of ligands

The surfaces of cells are often labelled with fluorescently tagged antibodies. However, cells can be incubated with a variety of labelled ligands. If analysis is performed in the presence of different concentrations of ligand, the surface density of the receptors and an affinity constant for binding can be calculated (Fay *et al.*, 1994; Sklar and Finney, 1982). Cells can also be stained simultaneously with an antibody to a surface marker and a labelled ligand (Loke *et al.*, 1992).

Receptor–ligand complexes often undergo endocytosis. Internalization of a fluorescently labelled ligand can be demonstrated by treating the cells with trypsin; ligand on the surface of the cell will be destroyed while internal ligand is protected. Flow cytometry can also be used to determine whether the internalized ligand enters a neutral or acidic compartment (lysosomes are acidic). If the ligand is dual labelled with a pH-sensitive dye (for example, fluorescein) and a pH-insensitive dye (rhodamine), the ratio of the fluorescein to rhodamine fluorescence indicates the pH of the environment of the ligand (Murphy *et al.*, 1984). The application requires two lasers, an argon-ion laser to excite the fluorescein and a krypton or a dye laser to excite the rhodamine.

This topic has been reviewed recently by Murphy (1990).

7.9 Intracellular calcium ions

Calcium ions have an important role in cell signalling and the intracellular concentration of calcium ions may show transient changes in response to external stimuli. There are several fluorescent dyes whose properties depend on the amount of bound Ca^{2+}. For flow cytometry, the most useful of these is indo-1 whose emission wavelength changes on binding calcium (see *Figure 3.3*). It is loaded into the cell in the form of the acetoxymethyl ester. An example of the type of data obtained is shown in *Figure 7.1*.

Although the measurement of qualitative changes in the concentration of intracellular Ca^{2+} is straightforward, quantitative measurements are more difficult. Calibration of the system can be a problem (see discussion by Rabinovitch and June, 1994).

7.10 Intracellular pH

This measurement is best made with a dye whose wavelength of emission is pH dependent. Molecular Probes have produced a series of benzo[c]xanthene dyes with these properties. The compound referred to as carboxy-SNARF-1 is excited optimally at 514 nm but can also be excited at 488 nm. The ratio of the orange (580) to red (630) fluorescent emissions is used to measure pH (*Figure 7.2*).

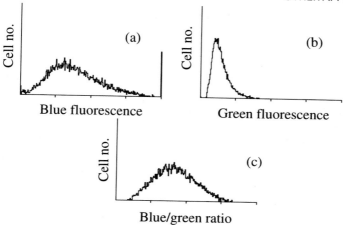

Figure 7.1:   Estimation of the concentration of intracellular calcium ions using indo-1. Cell line from human lymphoma (WIL2) labelled with 3 µM indo-1. (a) Histogram of blue fluorescence; (b) histogram of green fluorescence; (c) histogram of the ratio of the blue to green fluorescences. Ortho Cytofluorograf using an argon-ion laser tuned to produce UV. Cells prepared by Mrs L. Skelton, Institute of Cancer Research.

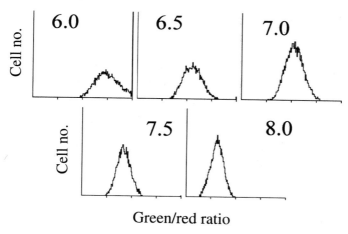

Figure 7.2:   Measurement of intracellular pH using SNARF-1. Cell line from human lymphoma (WIL2) labelled with 5 µM SNARF-1. Cells were suspended in buffers of different pH plus nigericin (2 µg ml^{-1}) to equalize buffer and intracellular pH. Green and red fluorescences were measured and the green/red ratio calculated for each cell. The numbers represent the pHs of the buffers used. As the pH increased, the green/red ratio decreased. Ortho Cytofluorograf. Excitation at 488 nm. Cells prepared by Mrs L. Skelton, Institute of Cancer Research.

7.11 Intracellular glutathione

There are non-fluorescent probes which react within the cells to give a fluorescent product (e.g. DCFH, see Section 7.6). An assay for intracellular glutathione is based on this approach. Monochlorobimane (MClB) crosses the plasma membrane and reacts with glutathione through glutathione-S-transferase to give a fluorescent conjugate which is trapped in the cell. MClB fluoresces blue when excited by UV or deep violet light. Watson (1990) has discussed the kinetics of measurements using MClB.

7.12 Chromosome analysis and sorting

Sorting individual chromosomes is an important application of flow cytometry. The DNA from a pure preparation of a particular chromosome can be amplified by a polymerase chain reaction and a reagent produced for 'painting' chromosomes by fluorescence *in situ* hybridization (FISH). Such paints from normal cells can be used to identify the origin of abnormal chromosomes. Alternatively, the same result can be achieved by applying a paint generated from an aberrant chromosome to a normal metaphase spread. These techniques are finding increasing application in the study of the genetics of human disease.

Unfortunately, chromosome sorting is not to be undertaken lightly. A suspension of chromosomes of good quality is essential (suitable protocols are to be found in Darzynkiewicz and Crissman, 1990; Monard and Young, 1994). Alignment of the instrument has to be undertaken with great care in order to maximize the resolution of the chromosomes. The cv of the chromosome stain should be less than 2% and, to achieve this, a laser power of several hundred milliwatts is required. Even then, a single DNA stain which will give fluorescence proportional to the DNA content (PI, for example) will not resolve the chromosomes sufficiently. To achieve better resolution, a dual stain is used to exploit variations in the GC to AT ratio in different chromosomes. The dyes usually selected are Hoechst 33258, which shows a preference for AT-rich regions, and chromomycin A_3, which binds to GC-rich regions. A cytometer with two high-powered argon-ion lasers is required; one laser is tuned to give UV to excite the Hoechst dye (blue fluorescence) and the other to 458 nm to excite the chromomycin A_3 (green fluorescence). A typical result is shown in *Figure 7.3*. Acceptable separation between the different chromosomes is achieved with the exception of chromosomes 9–12. The only way to separate these chromosomes is to use human–hamster hybrids in which only the desired chromosome amongst 9–12 is present.

The different aspects of chromosome analysis and sorting have been discussed by Gray and Cram (1990). Practical protocols have been given by Monard and Young (1994).

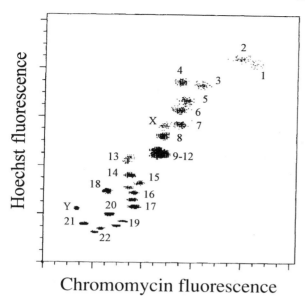

Figure 7.3:   A bivariate flow karyotype of a normal human male. The chromosomes were stained with Hoechst 33258 and chromomycin and the dyes excited with UV and at 458 nm. Heteromorphism of chromosomes 15, 16, 19 and 22 can be observed. Data recorded on a Becton-Dickinson FACSTAR Plus by Dr N.P. Carter, Department of Pathology, University of Cambridge, from chromosomes prepared during the Royal Microscopical Society's Advanced Flow Cytometry Course.

7.13 Tracking cells *in vivo*

Fluorescent dyes are a safe and convenient alternative to radioactive labels for following the migration of cells *in vivo*, particularly lymphocytes. The labelled cells are injected into an animal and subsequently are collected (e.g. by cannulation of a lymphatic duct) and counted in the flow cytometer. If required, the harvested cells can be labelled additionally with antibodies so that different populations of cells may be distinguished.

The surfaces of cells can be labelled covalently with fluorescein or rhodamine isothiocyanate, but the attachment of new groups to the surface proteins might alter the properties of the cells being studied. Horan and co-workers have reported the use of an analogue of acridine orange with an *N*-linked 26-carbon alkyl chain (Horan *et al.*, 1990). This dye, which is available commercially from Zynakis Cell Science, labels the lipids of the plasma membrane and remains in the membrane for up to 60 days. An alternative is to use the dye, Hoechst 33342, which is taken up by living cells, binds specifically to the DNA and persists in the nucleus for several days *in vivo* (Brennan and Parish, 1984; Ormerod, 1994).

7.14 Monitoring electropermeabilization

Electroporation is an efficient way of introducing foreign DNA into cells. The cells are permeabilized at a low temperature by a high voltage pulse and are then warmed to allow their membranes to reseal.

Several parameters, such as the medium used, the pulse voltage and the length and number of pulses, require optimization. Cells also vary in their electrosensitivity. Once the other parameters have been established, it is the pulse voltage which is dependent on the type of cell used. A flow cytometric procedure will establish whether holes have been punched in the plasma membrane and whether they have resealed successfully (Ormerod, 1994). By also using a simple assay of cell survival, the optimum voltage can be determined (O'Hare et al., 1989).

The method is a variant of that used to estimate cell viability (Section 6.4.1). PI is added immediately after electropermeabilization so that cells whose membranes have been ruptured take up the dye and their nuclei fluoresce red. After the cells have been incubated in warm medium to allow their membranes to reseal, FDA is added so that the cells with intact membranes (either unpermeabilized or whose membranes have successfully resealed) fluoresce green. The cells are excited at 488 nm and green vs. red fluorescence is recorded. Three clusters can be identified – green positive, red negative (cells not permeabilized); green positive, red positive (cells permeabilized and resealed membrane); green negative, red positive (cells over-permeabilized and membrane failed to reseal) (*Figure 7.4*).

7.15 Monitoring fusion or clustering of cells

The cells under study need to be labelled with two fluorescent dyes which do not affect the properties of the cell, stay associated with the labelled cell and do not migrate on to other cells under the conditions of the experiment. The fluorescence emission spectra of the dyes must also be sufficiently far apart. Generally, the most satisfactory results have been obtained with lipophilic dyes which label the plasma membrane, for example fluorophores with a long alkyl side chain attached (such as octadecyl fluorescein or 3,3′-dioctadecylindocarbocyanine).

For a study of cell fusion, if a dual laser cytometer is available then a combination such as octadecyl aminofluorescein and octadecyl rhodamine 6G can be used (Ormerod, 1994). Similar labels have been used to follow cell electrofusion and also cell fusion induced by polyethylene glycol (Ormerod and O'Hare, 1989). Cells labelled with the octadecyl dyes grow satisfactorily when put back into culture so that these dyes can be used to select hybrids by cell sorting.

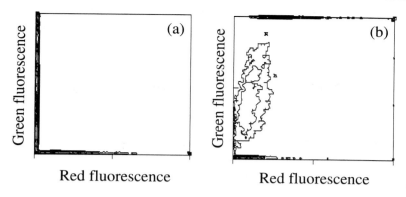

Figure 7.4: Monitoring electropermeabilization of a human neuroblastoma cell line (HX142). (a) No electropermeabilization; (b) after electropermeabilization; note the cluster of red +ve, green +ve cells. Cells prepared by Ms J. Mills and recorded by Mrs J. Titley, Institute of Cancer Research, on an Ortho Cytofluorograf. Excitation at 488 nm.

An alternative is to use two dialkyl cyanine dyes. Gant *et al.* (1992) employed 1,1'-dioctadecyl-3,3,3'3'-tetramethylindocarbocyanine and 3,3'-dioctadecyloxacarbocyanine to monitor the interaction between macrophages and T lymphocytes. Since the dyes are spectrally close, this combination is not ideal, but it has the advantage that a single laser can be used at 488 nm.

References

Assfalg-Machleid I, Rothe G, Klingel S, Banati R, Mangel WF, Valet G, Machleid W. (1992) Membrane permeable fluorogenic rhodamine substrates for selective determination of cathepsin L. *Biol. Chem. Hoppe-Seyler* **373**, 433–440.

Bass DA, Parce JW, Dechatelet LR, Szejda P, Seeds MC, Thomas M. (1983) Flow cytometric studies of oxidative product formation by neutrophils: a graded response to membrane stimulation. *J. Immunol.* **130**, 1910–1917.

Brennan M, Parish CR. (1984) Intracellular fluorescent labelling of cells for analysis of lymphocyte migration. *J. Immunol. Methods* **74**, 31–38.

Chused TM, Wilson HA, Seligmann BE, Tsien RY. (1986) Probes for use in the study of leukocyte physiology by flow cytometry. In *Applications of Fluorescence in the Biomedical Sciences* (eds DL Taylor, AS Waggoner, RF Murphy, F Lanni, R Birge). Alan R. Liss, New York, pp. 531–544.

Crissman HA, Darzynkiewicz Z, Tobey RA, Steinkamp JA. (1985) Correlated measurements of DNA, RNA and protein in individual cells by flow cytometry. *Science* **228**, 1321–1324.

Darzynkiewicz Z, Crissman HA. (eds) (1990) *Flow Cytometry Methods in Cell Biology*, Vol. 33. Academic Press, New York.

Darzynkiewicz Z, Kapuscinski J. (1990) Acridine orange: a versatile probe of nucleic acids and other cell constituents. In *Flow Cytometry and Sorting*, 2nd edn. (eds MR Melamed, T Lindmo, MI Mendelsohn). Wiley-Liss, New York, pp. 291–314.

Dyson JED, Britten RA, Battersby I, Surrey CR. (1989) Fluorescein isothiocyanate staining intensity as a probe of hyperthermia-induced changes in chromatin conformation. *Cytometry* **10**, 174–184.

Fay SP, Habbersett R, Domalewski MD, Posner RG, Houghton TG, Pierson E, Muthukumaraswamy N, Whitaker J, Haugland RP, Freer RJ, Sklar LA. (1994) Multiparameter flow cytometric analysis of a pH sensitive formyl peptide with application to receptor structure and processing kinetics. *Cytometry* **15**, 148–153.

Fiering SN, Roederer M, Nolan GP, Micklem DR, Parks DR, Herzenberg LA. (1991) Improved FACS-Gal: flow cytometric analysis and sorting of viable eukaryotic cells expressing reported gene constructs. *Cytometry* **12**, 291–301.

Gant VA, Shakoor Z, Hamblin AS. (1992) A new method of measuring clustering in suspension between accessory cells and T lymphocytes. *J. Immunol. Methods* **156**, 179–189.

Gray JW, Cram LS. (1990) Flow karyotyping and cell sorting. In *Flow Cytometry and Sorting*, 2nd edn. (eds MR Melamed, T Lindmo, MI Mendelsohn). Wiley-Liss, New York, pp. 503–529.

Horan PK, Melnicoff MJ, Jensen BD, Slezak SE. (1990) Fluorescent cell labelling for *in vivo* and *in vitro* cell tracking. In *Methods in Cell Biology*, Vol. 33 (eds Z Darzynkiewicz, HA Crissman). Academic Press, New York, pp. 469–490.

Krishan A. (1987) Effect of drug efflux blockers on vital staining of cellular DNA with Hoechst 33342. *Cytometry* **8**, 642–645.

Krishan A, Sridhar KS, Davila E, Vogel C, Sternheim W. (1987) Patterns of anthracycline retention modulation in human tumor cells. *Cytometry* **8**, 306–314.

Lee LG, Chen CH, Chiu LA. (1986) A new dye for reticulocyte analysis. *Cytometry* **7**, 508–517.

Loke YW, King A, Gardner L, Carter NP. (1992) Evidence for the expression of granulocyte-macrophage colony-stimulating factor receptors by human first trimester extravillous trophoblast and its response to cytokine. *J. Reprod. Immunol.* **22**, 33–45.

Monard SP, Young BD. (1994) Chromosome analysis and sorting by flow cytometry. In *Flow Cytometry. A Practical Approach*, 2nd edn. (ed. MG Ormerod). IRL Press, Oxford, pp. 169–183.

Murphy RF. (1990) Ligand binding, endocytosis, and processing. In *Flow Cytometry and Sorting*, 2nd edn. (eds MR Melamed, T Lindmo, MI Mendelsohn). Wiley-Liss, New York, pp. 352–366.

Murphy RF, Powers S, Cantor CR. (1984) Endosome pH measured in single cells by dual fluorescence flow cytometry: rapid acidification of insulin to pH 6. *J. Cell Biol.* **98**, 1757–1762.

Nolan GP, Fiering S, Nicolas JF, Herzenberg LA. (1988) Fluorescence-activated cell analysis and sorting of viable mammalian cells based on β-D-galactosidase activity after transduction of *Escherichia coli lacZ*. *Proc. Natl Acad. Sci. USA* **85**, 2603–2607.

O'Hare MJ, Ormerod MG, Imrie PR, Peacock JH, Asche A. (1989). Electropermeabilisation and electrosensitivity of different types of mammalian cells. In *Electroporation and Electrofusion in Cell Biology* (eds E Neumann, AE Sowers C Joran). Plenum Press, New York, pp. 319–330.

Ormerod MG. (1994) Further applications to cell biology. In *Flow Cytometry. A Practical Approach*, 2nd edn. (ed. MG Ormerod). IRL Press, Oxford, pp. 261–273.

Ormerod MG, O'Hare MJ. (1989) Monitoring cell fusion using flow cytometry. *Protides Biol. Fluids* **36**, 341–344.

Pollack A. (1990) Flow cytometric cell-kinetic analysis by simultaneously staining nuclei with propidium iodide and fluorescein isothiocyanate. In *Flow Cytometry. Methods in Cell Biology*, Vol. 33. (eds Z Darzynkiewicz, HA Crissman). Academic Press, San Diego, pp. 315–323.

Pollack A, Prudhomme DL, Greenstein DB, Irvin GL III, Claflin AJ, Block NL. (1982) Flow cytometric analysis of RNA content in different cell populations using pyronin Y and methyl green. *Cytometry* **3**, 28–35.

Rabinovitch PS, June CH. (1994) Intracellular ionised calcium, magnesium, membrane potential and pH. In *Flow Cytometry. A Practical Approach*, 2nd edn. (ed. MG Ormerod). IRL Press, Oxford, pp. 185–215.

Rothe G, Klingel S, Assfalg-Machleid I, Machleid W, Zirkerbach C, Banati R, Mangel WF, Valet G. (1992) Flow cytometric analysis of protein activities in vital cells. *Biol. Chem. Hoppe-Seyler* **373**, 547–554.

Sklar LA, Finney DA. (1982) Analysis of ligand-receptor interactions with the fluorescence activated cell sorter. *Cytometry* **3**, 161–165.

Waggoner AS. (1985) Dye probes of cell organelle and vesicle membrane potentials. In *The Enzymes of Biological Membranes* (ed. A Martonosi). Plenum Press, New York, pp. 313–331.

Watson JV. (1991) *Introduction to Flow Cytometry*. Cambridge University Press, Cambridge.

Weiss MJ, Chen LB. (1984) Rhodamine 123: a lipophilic mitochondrial-specific vital dye. *Kodak Lab. Chem. Bull.* **55**, 1–4.

Appendix A

Glossary

Analogue-to-digital converter: an electronic chip which changes analogue (continuously variable) signals to digital (discontinuous, binary) signals. In general, electronic circuitry (for example, an amplifier) uses analogue signals, computers work on digital signals.

Aneuploid: cytogeneticists use the word to describe a cell with an abnormal number of chromosomes. In flow cytometry, aneuploid is used to describe a cell whose DNA content is abnormal.

Autofluorescence: the fluorescence from an unlabelled cell. Cells contain substances which are fluorescent. The amount of autofluorescence observed will depend on the type of cell and the wavelength of excitation. Autofluorescence can limit the sensitivity of immunofluorescent detection.

Back gating: light scatter is often used to select a particular population of cells by setting a 'gate' (see below) on a cytogram of 90° vs. forward light scatter. If the desired population is not clearly apparent on the cytogram of light scatter, a fluorescently labelled antibody is used to pick out the cells; a gate is set for positive fluorescence and the light scatter of these cells displayed. A region can now be set on the light scatter of the gated cells and used to define these cells on an ungated cytogram of light scatter (see Section 4.2).

Blocker bar: *see* Obscuration bar.

Break-off point: in a cell sorter, the point at which the stream breaks up to form a droplet (see Section 2.8).

Coefficient of variation (cv): a dimensionless property which measures the spread of a population distribution (see Section 2.7.4).

Colour compensation: a correction factor applied to allow for spectral overlap (see below).

Contour plot: a display of a cytogram in which the density of cells is defined by contours (similar to those used on a cartographic map).

Cytogram: a two-dimensional histogram in which two cell parameters are correlated.

Deflection plates: in a cell sorter, two plates with a high voltage applied to them (typically 5000 V). The stream of droplets passes between the plates, charged droplets are deflected.

Dichroic mirror: an optical interference filter which reflects one colour and transmits another. These filters either transmit longer wavelengths and reflect shorter (long pass) or vice versa (short pass).

Diploid: cytogeneticists use the word to describe a cell with the normal number of chromosomes. In flow cytometry, diploid is used to describe a cell whose DNA content is normal.

DNA index (DI): the DNA content of tumour cells in G1 of the cell cycle compared to that of normal cells, hence normal cells have a DI of 1.

Dot plot: a representation of a cytogram in which each cell is represented by a dot on the two-dimensional graph.

Droplet delay: in a cell sorter, the time difference between a cell crossing the laser beam and a droplet, containing that cell, being formed (see Section 2.8).

FCS format: Flow Cytometry Standard format for files of data. The standard is published by the International Society for Analytical Cytology in its journal, *Cytometry*. FCS 2.0 is used by the major manufacturers and can be read by all third party software.

Flow karyotype: the pattern of chromosomes as revealed by analysis in a flow cytometer (see Section 7.12).

Gate: a gate is selected by defining a region on a univariate histogram or a cytogram (bivariate histogram). Only cells falling within the gate can pass through to the next stage of analysis (see Section 2.7.2). Gates are also used to select desired populations for cell sorting.

Interrogation point: the point in the sample stream at which the laser light is focussed. At this point the cells are measured, or interrogated.

Isometric plot: a pseudo-three-dimensional representation of a two-dimensional cytogram in which cell number is shown on the third (vertical) axis.

Jet-in-air (JIA): *see* Stream-in-air.

Linear amplifier: an electronic amplifier whose output signal is proportional to the input signal. *See also* Logarithmic amplifier.

Logarithmic amplifier: an electronic amplifier whose output signal is proportional to the logarithm of the input signal.

Obscuration bar: a bar positioned in front of a collection lens to block laser light. In a forward direction, an obscuration bar is used to block the primary laser beam after it has passed through the flow cell or sample stream. In 'stream-in-air' instruments, an obscuration bar is also placed in front of the lens which collects 90° scatter and fluorescent light.

Pulse shape analysis: when a cell passes through the laser beam, scattered or fluorescent light is detected and an electronic signal (pulse) is generated. Analysis of the shape of the pulse can give information about the size and shape of the cell (see Section 5.5).

Sheath fluid: the fluid which passes through the flow cell. The sample is injected into the stream of sheath fluid.

Spectral overlap: the fluorescence emission spectrum from one fluorochrome may overlap the emission spectrum of another. For example, the emission spectra of fluorescein and phycoerythrin overlap and cannot be completed separated. *See also* Colour compensation.

Stream-in-air: this phrase refers to the arrangement of the flow cell in a cell sorter. The laser beam is focussed in the stream after it has emerged from the flow cell. Sometimes called 'jet-in-air' (JIA).

Tetraploid: describes a cell with double the DNA content of a normal (diploid) cell.

Threshold: the electronic circuitry is set to accept information (that is, to 'trigger') only when the input signal exceeds a set level (the threshold). The threshold is usually set on light scatter. Too low a threshold allows the system to accept too much noise, at too high a level some small particles may be excluded.

Appendix B

Manufacturers and suppliers

Manufacturers of flow cytometers

Becton-Dickinson, Immunocytometry Systems, 2350 Qume Drive, San José, CA 95131-1807, USA.

Becton-Dickinson UK, Between Towns Road, Cowley, Oxford OX4 3LY, UK.

Biorad Laboratories Ltd, Caxton Way, Watford Business Park, Watford, Herts WD1 8RP, UK.

Coulter Electronics Ltd, Northwell Drive, Luton, Beds LU3 3RH, UK.

Ortho Diagnostic Systems Ltd, Enterprise House, Station Road, Loudwater, High Wycombe, Bucks HP10 9UF, UK, *and* Route 202, Raritan, NJ 08869, USA.

Partec GmbH, Hüfferstrasse 73–79, D-4400 Münster, Germany.

Suppliers of optical filters

Glen Spectra Ltd, 2–4 Wigton Gardens, Stanmore, Middlesex HA7 1BF, UK.

Melles Griot, 1770 Kettering Street, Irvine, CA 92714, USA.

Oriel Corporation, 250 Long Beach Blvd, PO Box 872, Stratford, CT, USA.

The Melles Griot catalogue contains helpful information about the design and construction of filters.

Suppliers of fluorescent dyes

Exciton Inc., PO Box 31126, Overlook Station, Dayton, OH 45431, USA.

Molecular Probes, Inc., 4849 Pitchford Avenue, Eugene, OR 97402, USA.

The Molecular Probes catalogue contains a wealth of information about hundreds of different dyes, together with an extensive reference list.

Zynakis Cell Science Inc., 371 Phoenixville Pike, Malvern, PA 19355, USA.

Lasers

Coherent Inc., 3210 Porter Drive, PO Box 1032, Palo Alto, CA 94304, USA.

Spectra-Physics, Boundary Way, Hemel Hempstead, Herts HP2 7SH, UK, *and* 1250W Middlefield Road, PO Box 7013, Mountain View, CA 94039-7013, USA.

Laser servicing in the UK

The laser manufacturers (see above) offer laser servicing. There are also several independent companies.

Cambridge Lasers Ltd, Brookfield Business Centre, Cottenham, Cambridge CB4 4PS.

Laser Support Services, 38b James Street, Pittenweem, Fife KY10 2QN.

Antibodies

Amersham International, Lincoln Place, Green End, Aylesbury, Bucks HP20 2TP, UK.

Becton-Dickinson Immunocytometry Systems, address above.

Caltag Laboratories, 436 Rozzi Place, South San Francisco, CA 94080, USA.

Coulter Electronics, address above.

Dako A/S, 42 Produktionsvej, DK-2600, Glostrup, Denmark.

Immunotech S.A., Luminy-Case 915, 13288 Marseille, Cedex 9, France.

Ortho Diagnostics, address above.

Pharmingen, 11555 Sorrento Valley Road, San Diego, CA 92121, USA.

Sera-Lab Ltd, Crawley Down, Sussex RG10 4FF, UK.

Tago Inc., PO Box 4463, Burlingame, CA 94011, USA.

Fluorescent standards, quantification and fluorescent beads

Biocytex, 140 Chemin de l'Armée d'Afrique, 13010 Marseille, France.

Flow Cytometry Standards Corp., PO Box 194344, San Juan, Puerto Rico 00919.

Polysciences, Inc., 400 Valley Road, Warrington, PA 18976-2590, USA.

Software

Dako A/S, address above.

Phoenix Flow Systems, 11575 Sorrento Valley Road, San Diego, CA 92121, USA.

Verity Software House Inc., 10 New Lewiston Road, PO Box 247, Topsham, ME 04086, USA.

Appendix C

Learned societies

The Royal Microscopical Society (37/38 St Clements, Oxford OX4 1AJ, UK; fax: (44) (0)865 791237) has a cytometry section which caters for people interested in flow cytometry. It organises meetings, workshops and courses.

The International Society for Analytical Cytology (PO Box 7849, Breckenridge, CO 80424-7849, USA; fax: (1)303 453 2636) organises a large symposium on image and flow cytometry every 18 months. The membership fee includes a subscription to their journal, *Cytometry*. Almost all of the important new developments in flow cytometry are published in this journal. A preferential subscription may also be obtained to *Communications in Clinical Cytometry*.

The European Society for Analytical Cellular Pathology (apply to Elsevier Scientific Publishers Ireland, Ltd, Bay 15, Shannon Industrial Estate, Shannon, Co. Clare, Ireland; fax: (6)472144) organises a bi-annual meeting on image and flow cytometry. The membership fee includes a subscription to their journal, *Analytical Cellular Pathology*. At present, this journal is biased towards clinical applications of image cytometry.

Index

OTHER MICROSCOPY HANDBOOKS

Confocal Laser Scanning
Microscopy
C. Sheppard & D. Shotton

Food Microscopy
O. Flint

Enzyme Histochemistry
A Laboratory Manual of Current
Methods
C. J. F. van Noorden &
W. M. Frederiks

The Role of Microscopy in
Semiconductor Failure Analysis
B. P. Richards & P. K. Footner

Qualitative Polarized-Light
Microscopy
P. C. Robinson & S. Bradbury

The Preparation of Thin
Sections of Rocks, Minerals
and Ceramics
D. W. Humphries

Introduction to Crystallography
Revised Edition
C. Hammond

Basic Measurement Techniques
for Light Microscopy
S. Bradbury

An Introduction to Surface
Analysis Electron Spectroscopy
J. F. Watts

Cryopreparation of Thin
Biological Specimens for
Electron Microscopy
Methods and Applications
N. Roos & A. J. Morgan

The Operation of Transmission
and Scanning Electron
Microscopes
D. Chescoe & P. J. Goodhew

Autoradiography
A Comprehensive Overview
J. R. J. Baker

RMS Dictionary of Light
Microscopy
Compiled by the Nomenclature
Committee of the RMS

An Introduction to the Optical
Microscope
Revised Edition
S. Bradbury

Colloidal Gold
A New Perspective for Cytochemical
Marking
J. E. Beesley

Light-Element Analysis in the
Transmission Electron
Microscope
WEDX and EELS
P. M. Budd & P. J. Goodhew

An Introduction to Scanning
Acoustic Microscopy
A. Briggs

An Introduction to
Immunocytochemistry
Current Techniques and Problems
Revised Edition
J. M. Polak & S. van Noorden

Maintaining and Monitoring the
Transmission Electron
Microscope
S. K. Chapman

X-Ray Microanalysis in Electron
Microscopy for Biologists
A. J. Morgan

Lipid Histochemistry
O. Bayliss High

IN SITU HYBRIDIZATION

A. R. Leitch, T. Schwarzacher, D. Jackson & I. J. Leitch
respectively Queen Mary and Westfield College, London, UK; John Innes Research Centre, Norwich, UK; USDA, Plant Gene Expression Center, Albany, California, USA; and Royal Botanic Gardens, Kew, UK

In situ hybridization is a powerful link between cellular and molecular biology. This new practical guide provides a comprehensive description of *in situ* hybridization, from background information to detailed methodology and practical applications. Its clarity of approach and up-to-date coverage of methods and troubleshooting make it the ideal introduction for all first-time users of *in situ* hybridization and a valuable companion for established researchers.

Contents
- Introduction
- Nucleic acid sequences located *in situ*
- The material
- Nucleic acid probes, labels and labelling methods
- Denaturation, hybridization and washing
- Detection of the *in situ* hybridization sites
- Imaging systems and the analysis of signal
- The *in situ* hybridization schedule (including troubleshooting)
- The future of *in situ* hybridization

Suitable for third-year undergraduate and postgraduate students of molecular and cell biology and genetics; invaluable to both first-time users and experienced researchers.

Microscopy Handbook, No. 27
Paperback, 128 pp, 2 colour plates, 1994
1 872748 48 1

BIOLOGICAL MICROTECHNIQUE

J. Sanderson
Sir William Dunn School of Pathology, Oxford, UK

Although many significant advances have been made in biological specimen preparation during the past twenty years, no new practical guide to the techniques has been published in this time. As a result of the recent resurgence of interest in light microscopy, particularly confocal techniques, this new, up-to-date book will benefit both novices and experienced microscopists seeking to extend their repertoire of techniques.

A poorly-prepared specimen inevitably leads to unreliable results. This new book therefore describes both new and classical methods of slide-making in an easy-to-read, easy-to-understand format. It contains a wealth of practical detail which will provide a firm grounding in preparative methods for light microscopy.

Contents include
- Fixation
- Tissue processing
- Microtomy
- Other preparative methods
- Staining and dyeing
- Finishing the preparation

Suitable for junior researchers; laboratory technicians; university students; skilled amateur microscopists; school science teachers.

Microscopy Handbook, No. 28
Paperback, 240 pp, 1994
1 872748 42 2

LIGHT MICROSCOPY

D. J. Rawlins
John Innes Institute, Norwich, UK

". . . remarkably comprehensive and full of practical tips about how to choose a microscope or get more out of your old one" *Bulletin of the Royal College of Pathologists*

"One of the best introductions to practical research light microscopy that is currently available" *Choice*

Contents include: Choice of imaging method, Specimen preparation, Setting up different imaging methods, Trouble-shooting, Three-dimensional microscopy, Taking photomicrographs, and Video microscopy.

Introduction to Biotechniques series
Paperback, 156 pp, 1992
1 872748 11 2

LIGHT MICROSCOPY
An Electronic Textbook

Script by D. J. Rawlins
Produced by AVC Multimedia Ltd for the COMETT/Biotechnology in Training (BIT) Consortium

"It could definitely serve as a self-learning 'overview' or 'reinforcing' tutorial that students could access in their own time. Possibly its greatest value lies in the ability of the program to give a rapid insight into the range of techniques available . . . as well as aspects of their practical use." *CABIOS*

This easy-to-use training package is the ideal way to teach more students for less cost! The package includes high definition images (many in full colour), menu facility, electronic bookmark and "map" functions to guide the student through: What sort of microscopy to use, Basic principles of the microscope and magnification, Techniques of photomicrography and measuring, Bright-field, Phase contrast, Fluorescence, Dark-field, Polarized light, Nomarski, and Reflected light microscopy.

System requirements: IBM pc or compatible; minimum of 286 machine but 386 with SVGA recommended, 1 MB video RAM and a 40 MB hard disk.
Single installation 1 872748 31 7
Department installation 1 872748 97 X

ORDERING DETAILS

Main address for orders

BIOS Scientific Publishers Ltd
St Thomas House, Becket Street,
Oxford OX1 1SJ, UK
Tel: +44 865 726826
Fax: +44 865 246823

Australia and New Zealand
DA Information Services
648 Whitehorse Road, Mitcham, Victoria 3132, Australia
Tel: (03) 873 4411
Fax: (03) 873 5679

India
Viva Books Private Ltd
4346/4C Ansari Road, New Delhi 110 002, India
Tel: 11 3283121
Fax: 11 3267224

Singapore and South East Asia
(Brunei, Hong Kong, Indonesia, Korea, Malaysia, the Philippines,
Singapore, Taiwan, and Thailand)
Toppan Company (S) PTE Ltd
38 Liu Fang Road, Jurong, Singapore 2262
Tel: (265) 6666
Fax: (261) 7875

USA and Canada
Books International Inc
PO Box 605, Herndon, VA 22070, USA
Tel: (703) 435 7064
Fax: (703) 689 0660

Payment can be made by cheque or credit card (Visa/Mastercard, quoting number and expiry date). Alternatively, a *pro forma* invoice can be sent.

Prepaid orders must include £2.50/US$5.00 to cover postage and packing for one item and £1.25/US$2.50 for each additional item.